D1063962

DISCRETE-TIME SYSTEMS
An Introduction to the Theory

Discrete-Time Systems

AN INTRODUCTION TO THE THEORY

Herbert Freeman

Professor of Electrical Engineering

New York University

JOHN WILEY & SONS, INC.

New York · London · Sydney

Library of Congress Catalog Card Number: 65-14255
Printed in the United States of America

To My Parents

Since the early 1950's the increasing use of digital computers has been quietly causing a major revolution in the fields of engineering and science. Originally developed merely as an aid for time-consuming numerical calculations, the digital computer has today grown into a valuable research assistant, capable not only of solving problems but even of suggesting them. The rapidity with which a computer can perform arithmetic operations and make decisions has caused us to seek entirely new approaches to formulating mathematical models. Complexity per se is no longer the great obstacle that it once was. Instead of being limited to simple mathematical models, easily amenable to paper-and-pencil study, we now feel free to study problems with hundreds of variables, to explore all kinds of "secondary" effects, and to perform calculations to precisions that exceed our measuring abilities. Particularly in the field of system theory—that is, the study of the mathematical models of *dynamic* phenomena—the digital computer has enormously increased the variety of problems that can be usefully studied as well as the depths to which the subtleties of these problems can be explored.

Digital computers perform their calculations discretely in time. Hence the use of a digital computer for system study, whether for direct simulation or merely indirect analysis, requires that time be a discrete variable. A system under study on a computer thus necessarily becomes a *discrete-time system.*

The replacement of a continuous-time system by a discrete-time system must be carried out with caution and understanding if meaningful results for the continuous-time system are to be obtained from an analysis of the discrete-time system. The techniques for studying discrete-time systems differ from those of continuous-time systems. However, relatively few books in science and engineering have been directly concerned with the study of discrete-time systems. There are many excellent books on sampled-data control systems, on the calculus of finite differences, and on numerical analysis. It is my belief, however, that these books are addressed to more specialized topics and do not sufficiently emphasize discrete-time theory as a subject in its own right. It was for this reason that this book was written.

For a number of years I have been conducting a course on the subject of discrete-time system theory at New York University. The lecture notes used in this course and distributed to the students have, after many revisions, resulted in this book. The book is intended primarily as a textbook for a one-semester graduate course for engineering students. The course can be either an independent course in discrete-time systems or part of a course sequence in control systems or general system theory.

In presenting the material in this book, I have taken care to minimize the background information expected from the reader, whether graduate student or practicing engineer. The reader will encounter no difficulty in mastering the subject matter if he is familiar with the elementary principles of the Laplace and Fourier transforms, linear system theory, complex variables through contour integration, and matrix algebra. Some basic knowledge in difference equations, control theory, and statistics would be helpful but is not required. It has been my experience that the variations in background, even of students recently graduated in the same engineering discipline, tend to be so extreme that the assumption of a more advanced background would severely restrict the book's usefulness.

The book is organized into eight chapters. The first chapter defines the basic terms used in system theory and introduces the modern concept of system description based on the use of state variables.

The second chapter is concerned with time-domain analysis of linear, discrete-time systems. Both the convolution-sequence and the state-variable approaches are discussed and related to each other. Emphasis is placed on the use of state variable techniques. Included are discussions of state-space transformations and the concepts of controllability and observability.

Chapter 3 is devoted entirely to transformation calculus techniques for linear, stationary systems. The emphasis is on the so-called two-sided transformation. The mathematically better-known term of *generating function* is employed for the general discrete-time transformation, and the term *z transform* is reserved for the generating function of infinite sequences that are nonzero only for positive time.

In Chapter 4 the problem of sampling a continuous-time function is examined in detail. The sampling theorem is derived and extended to a variety of special cases.

Chapter 5 is concerned with the interpolation and extrapolation of sampled data by means of approximating functions. The relation between various classical interpolation methods is discussed. Emphasis is placed on the so-called hold extrapolators, so important in sampled-data control systems.

Chapter 6 is devoted to methods for analyzing continuous-time systems that are subjected to discrete-time inputs.

The subject of Chapter 7 is sampled-data control systems. Methods based on the use of transfer functions as well as methods using the state-variable approach are described. The important problem of system stability is discussed from the viewpoint of Liapunov's method.

In Chapter 8 the material of the earlier chapters is applied to systems with stochastic signals as well as to finite-state, probabilistic systems. An illustration is given of the design of an optimum Markov system using the method of dynamic programming.

Two appendices are included, one describing a numerical technique for inverting rational z transform expressions, and the other consisting of tables of z transform pairs. A set of problems, arranged by chapters, is included at the end of the book. A list of major pertinent references is given at the end of each chapter.

During the process of writing this book, I received many valuable suggestions from colleagues, associates, students, and friends. They are too numerous for me to mention and thank them individually. There are a few, however, whom I should like to single out for special thanks: Dean John R. Ragazzini of New York University for originally arousing my interest in this field; Dean Gordon S. Brown of the Massachusetts Institute of Technology for offering me a visiting professorship there, during which time the idea of writing this book was first formed; Mr. Arthur A. Hauser of the Sperry Rand Research Center, Sudbury, Massachusetts, for many hours of stimulating discussions; and Dr. Rudolf F. Drenick of the Polytechnic Institute of Brooklyn for a host of helpful criticisms and comments. Finally, I owe a debt of gratitude to my wife Joan for her constant encouragement, without which this book would never have been completed.

HERBERT FREEMAN

December 1964
Great Neck, New York

Contents

CHAPTER 1

Basic System Concepts

1.1 INTRODUCTION

Engineers and physical scientists have for many years utilized the concept of a *system* to facilitate the study of the interaction between forces and matter. A system is a mathematical abstraction that is devised to serve as a model for a dynamic phenomenon. It represents the dynamic phenomenon in terms of mathematical relations among three sets of variables known as the *input*, the *output*, and the *state*.

The *input* represents, in the form of a set of time functions or sequences, the external forces that are acting upon the dynamic phenomenon. In similar form, the *output* represents the measures of the directly observable behavior of the phenomenon. Input and output bear a cause–effect relation to each other; however, depending on the nature of the phenomenon, this relation may be strong or weak.

A basic characteristic of any dynamic phenomenon is that the behavior at any time is traceable not only to the presently applied forces but also to those applied in the past. We may say that a dynamic phenomenon possesses a "memory" in which the effect of past applied forces is stored. In formulating a system model, the *state* of the system represents, as a vector function of time, the instantaneous content of the "cells" of this memory. Knowledge of the state at any time t_0 plus knowledge of the forces subsequently applied is sufficient to determine the output (and state) at any time $t \geq t_0$.

As an example, a set of moving particles can be represented by a system in which the state describes the instantaneous position and momentum of each particle. Knowledge of position and momentum, together with knowledge of the external forces acting on the particles (i.e., the system input) is sufficient to determine the position and momentum at any future time.

The general relations among the input, output, and state of a system are illustrated in Fig. 1-1. This representation is instructive from a conceptual point of view; however, we should note that the "memory" of a system is in general not as distinct as implied here but is likely to be distributed over many elements of the system.

A system is, of course, not limited to modeling only physical dynamic phenomena; the concept is equally applicable to abstract dynamic phenomena such as those encountered in economics or other social sciences.

One of the most fundamental principles of system theory is that of nonanticipation. A system is said to be *nonanticipatory* (or *physically realizable*) if its state and output at any time t_0 may be a function of only those input values that have occurred at $t < t_0$; that is, a nonanticipatory system does not respond to input values until after their occurrence. All systems in which "time" is true clock time must be nonanticipatory.

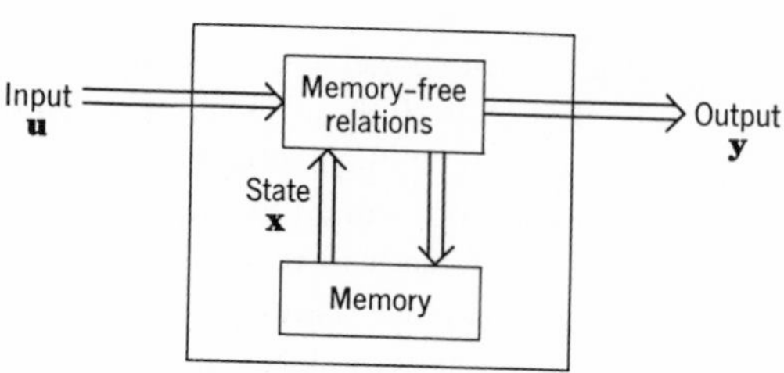

Figure 1-1 Conceptual representation of a system. The **u**, **x**, and **y** are vector time functions of dimensions m, n, and p, respectively.

A system is said to be *deterministic* if its state and output at any time t can be determined with certainty from a complete knowledge of its state at some time t_0 and its input over the semiclosed time interval $[t_0, t)$.[1] Conversely, a system is *stochastic* (or nondeterministic) if such knowledge of state and input suffices only to provide a statistical description of the state and output at time t.

Since the input, state, and output in general consist of sets of variables, we represent them by vector quantities; for example, an m-variable input is written as

$$\mathbf{u} = \begin{bmatrix} u_1 \\ u_2 \\ \cdot \\ \cdot \\ \cdot \\ u_m \end{bmatrix} = [u_i], \qquad i = 1, 2, \ldots, m \tag{1.1}$$

The state and output are denoted by the vectors **x** and **y** respectively, where **x** is of dimension n and **y** of dimension p. Different vectors of the

[1] This definition applies to a nonanticipatory system. For an anticipatory system, the time interval may be $[t_0, t_1)$, where $t_1 > t$.

same class are distinguished by means of superscripts, for example, $\mathbf{u}^1$, $\mathbf{u}^2$, etc.

The *input space* U represents the set of all possible inputs $\mathbf{u}$ of the system. Similarly, the *state space* X represents the set of all possible states of the system, and the *output space* Y, the set of all possible outputs. The set of all time values for which $\mathbf{u}$, $\mathbf{x}$, and $\mathbf{y}$ are defined is the *time space*[2] Θ.

If the time space is continuous, the system is known as a *continuous-time system*. However, if the input and state vectors are defined only for discrete instants of time t_k, where k ranges over the integers, the time space is discrete and the system is referred to as a *discrete-time system*.

We shall denote a function of continuous time by $\{f(t)\}$ and its value at t by $f(t)$. Similarly, a function of discrete time shall be denoted by $\{f(k)\}$ and its value at $t = t_k$ by $f(k)$. Vector functions will be in bold face. Where no ambiguity can arise, a function may be represented simply by f, or by $\mathbf{f}$ if it is a vector function. Discrete-time functions will also be referred to as *time sequences*.

It is important to distinguish between functions whose argument is discrete (i.e., functions of a discrete variable) and those that in themselves vary over a discrete set. Functions of the latter type will be referred to as *quantized* functions and the systems in which they appear will be called *quantized-data systems*. Thus the function illustrated in Fig. 1-2*a* is a discrete-time function and the one shown in Fig. 1-2*b* is a quantized function. Note that in Fig. 1-2*b* $\{f(t)\}$ can range only over discrete values (which need not necessarily be uniformly spaced). A discrete-time function may, of course, be quantized as well; this is shown in Fig. 1-2*c*.

In certain continuous-time systems, some state variables are allowed to change only at discrete instants of time t_k, where k ranges over the integers and where the spacing between successive instants may be arbitrary or uniform. Such systems are in effect a kind of hybrid between a continuous-time and a discrete-time system. They are encountered whenever a continuous-time function is sampled at discrete instants of time. Although they are frequently analyzed most easily by treating them as discrete-time systems, they differ from discrete-time systems in that special consideration may have to be given to the instants of time at which the sampling occurs. They have been given a special name and are known as *sampled-data systems*.

We shall be concerned here primarily with the analysis of discrete-time systems. Our interest in these systems is motivated by a desire to predict the performance of the physical devices for which this kind of system is an appropriate model. This is, however, not the only motive for studying

[2] The input, state, and output spaces are real and finite-dimensional, and Θ consists of the set (or a subset) of real numbers.

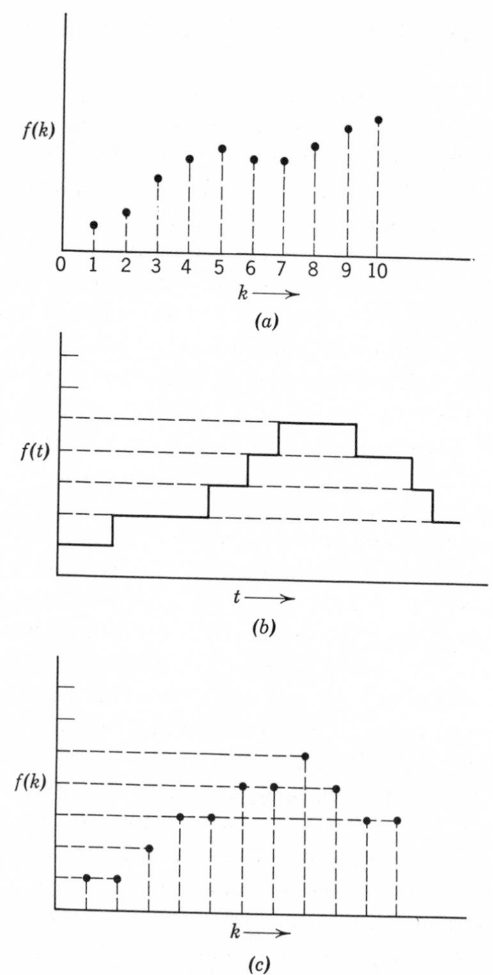

Figure 1-2 Illustration of discrete-time functions and quantized functions. (*a*) discrete-time function, (*b*) quantized function, (*c*) quantized, discrete-time function

discrete-time systems. A second, nearly as important reason is that there are many continuous-time systems that are more easily analyzed when a discrete-time model is fitted to them. A common example of this is the simulation of a continuous-time system by a digital computer. Further, a considerable body of mathematical theory has been developed for the analysis of discrete-time systems. Much of this is valuable for gaining insight into the theory of continuous-time systems as well as discrete-time systems. We note that a continuous-time function can always be viewed as the limit of a time sequence whose spacing between successive terms is allowed to become infinitesimal. Conversely, a time sequence can be obtained from a continuous-time function by a sampling operation.

1.2 STATE-SPACE REPRESENTATION

Consider a deterministic system[3] that is at rest prior to time t_0. This condition exists in a physical dynamic phenomenon when all the interacting objects are in their rest positions and possess zero momentum, that is, when both the potential and the kinetic energy are zero. An input $\mathbf{u}$ is applied to the system, beginning at t_0. The output $\mathbf{y}(t)$ at some time $t \geq t_0$ will be solely a function of $\mathbf{u}$ over the semiclosed time interval $[t_0, t)$; that is, for all $\mathbf{u}$ in U and all t_0, t in Θ,

$$\mathbf{y}(t) = G_1(\mathbf{u}_{[t_0, t)}) \tag{1.2}$$

We have defined the output of a system as the system's *directly observable behavior*. The output variables thus correspond to physically measurable quantities. It is conceivable that a system may in addition also possess behavior characteristics that are not directly observable. In accordance with our earlier definition we refer to the system's *total internal condition* or behavior as the *state* of the system and regard the output space Y as a subspace of X. If this viewpoint is taken, the state at time t of a system that is at rest at t_0, $t \geq t_0$, is solely a function of $\mathbf{u}$ over the semiclosed interval $[t_0, t)$, and the output is expressible simply as a function of the state at time t. Thus for all $\mathbf{x}$ in X, all $\mathbf{u}$ in U, and all t_0, t in Θ,

$$\mathbf{x}(t) = G_2(\mathbf{u}_{[t_0, t)}) \tag{1.3}$$

$$\mathbf{y}(t) = G_3(\mathbf{x}(t)) \tag{1.4}$$

If we now remove the requirement that the system be at rest at t_0,

$$\mathbf{x}(t) = G_4(\mathbf{x}(t_0); \mathbf{u}_{[t_0, t)}) \tag{1.5}$$

where $\mathbf{u}_{[t_0, t)}$ represents the input $\mathbf{u}$ over the semiclosed interval $[t_0, t)$, and the state $\mathbf{x}(t_0)$ reflects the effect on the system of all inputs for $t < t_0$.

Equations (1.4) and (1.5) are known as the *state equations* of the system; they completely characterize a deterministic system. The dimension n of the vector $\mathbf{x}$ is referred to as the *order* of the system. Note that from (1.5), for any time t_1, $t \geq t_1 \geq t_0$,

$$\mathbf{x}(t) = G_4(\mathbf{x}(t_1); \mathbf{u}_{[t_1, t)}) \tag{1.6}$$

$$= G_4[G_4(\mathbf{x}(t_0); \mathbf{u}_{[t_0, t_1)}); \mathbf{u}_{[t_1, t)}] \tag{1.7}$$

[3] Unless otherwise stated, a system is always assumed to be nonanticipatory.

If the system relations can be expressed in terms of differential equations, the state equations (1.4) and (1.5) may be written in the form

$$\frac{d}{dt}\mathbf{x}(t) = G_5\left(\mathbf{x}(t); \mathbf{u}(t); \frac{d}{dt}\mathbf{u}(t); \frac{d^2}{dt^2}\mathbf{u}(t); \ldots; \frac{d^v}{dt^v}\mathbf{u}(t)\right) \tag{1.8}$$

$$\mathbf{y}(t) = G_6(\mathbf{x}(t)) \tag{1.9}$$

The integer v in (1.8) depends on the particular form of the differential equations; for most commonly encountered systems it is equal to zero.

1.3 OPERATORS

In the study of systems, much use is made of the concept of operators. Operators serve primarily to simplify notation and thereby facilitate the formulation and solution of both analysis and synthesis problems. In addition they sometimes provide insight into the functional behavior of the physical devices they represent. We shall accordingly examine the concept of operators in some detail.

An *operator* is a transformation to be carried out on an operand. For generality we let the operand be an r-dimensional vector representing a set of r variables. The set of all operands for which the transformation is defined is called the *domain* of the operator. The corresponding set of results of the transformation is called the *range* of the operator. An operator is represented by a symbol which is placed to the left of that of the operand. The combination of operator and operand represents the result of the indicated transformation.

Given an operator H defined for a domain F and possessing a range Σ, we may write for an r-vector operand $\mathbf{f}$ contained in F:

$$\mathbf{y} = H\mathbf{f} \tag{1.10}$$

where $\mathbf{y}$ is contained in Σ and is also an r-vector. Equation (1.10) is to be interpreted in the sense that the operator H operates separately on each component of $\mathbf{f}$; that is,

$$y_i = Hf_i \quad \text{for all } i, \qquad i = 1, 2, \ldots, r \tag{1.11}$$

where f_i and y_i are corresponding components of the vectors $\mathbf{f}$ and $\mathbf{y}$, respectively.

Two operators H_1 and H_2 are equivalent if their domains and ranges are respectively equal and if

$$H_1\mathbf{f} = H_2\mathbf{f}$$

for all $\mathbf{f}$ in the common domain.

By definition, the operator product $H = H_2H_1$ when applied to an operand $\mathbf{f}$ represents the result of the transformation specified by the operator H_2 operating on the results of H_1 operating on $\mathbf{f}$; that is,

$$H\mathbf{f} = H_2H_1\mathbf{f} = H_2(H_1\mathbf{f}) \tag{1.12}$$

provided the range of H_1 lies within the domain of H_2. The operator product simply represents the combined, properly cascaded transformations of the operators H_2 and H_1.

Successive applications of the same operator may be simply denoted by the operator raised to a power; for example,

$$HH\mathbf{f} = H^2\mathbf{f} \tag{1.13}$$

It is to be noted that the operator product is in general not commutative; that is,

$$H_2H_1\mathbf{f} \neq H_1H_2\mathbf{f} \tag{1.14}$$

The sum of two operators H_1 and H_2 is defined by

$$(H_1 + H_2)\mathbf{f} = H_1\mathbf{f} + H_2\mathbf{f} \tag{1.15}$$

for all $\mathbf{f}$ within the domain common to the two operators. It follows that

$$(H_1 + H_2)\mathbf{f} = (H_2 + H_1)\mathbf{f} \tag{1.16}$$

and

$$[H_1 + (H_2 + H_3)]\mathbf{f} = [(H_1 + H_2) + H_3]\mathbf{f} \tag{1.17}$$

where $\mathbf{f}$ must now lie in the domain common to all three operators. The three relations (1.15). (1.16), and (1.17) represent, respectively, the distributive, commutative, and associative laws for operator summation.

The sum $\mathbf{b}$ of two r-vectors $\mathbf{f}$ and $\mathbf{g}$ is another r-vector, expressed by

$$\mathbf{b} = \mathbf{f} + \mathbf{g} \tag{1.18}$$

where

$$b_i = f_i + g_i \quad \text{for all } i, \qquad i = 1, 2, \ldots, r \tag{1.19}$$

Note that an operator is in general not distributive over operands; that is,

$$H(\mathbf{f} + \mathbf{g}) \neq H\mathbf{f} + H\mathbf{g} \tag{1.20}$$

even though both $\mathbf{f}$ and $\mathbf{g}$ may lie within the domain of H.

Multiplication of an r-vector by a scalar yields an r-vector,

$$\mathbf{b} = a\mathbf{f} \tag{1.21}$$

where

$$b_i = af_i \quad \text{for all } i \qquad i = 1, 2, \ldots, r \tag{1.22}$$

An operator is said to be *homogeneous* if

$$H a\mathbf{f} = aH\mathbf{f} \tag{1.23}$$

for any scalar a and all $\mathbf{f}$ in the domain of H.

The most general transformation is represented by the *vector operator* $\mathbf{H}(i)$. The expression

$$y_i = \mathbf{H}(i)\mathbf{f}, \qquad i = 1, 2, \ldots, s \tag{1.24}$$

is defined as the mapping of the r-vector $\mathbf{f}$ into the ith component of the s-vector $\mathbf{y}$.

An important special case of (1.24) is the *operator matrix*:

$$\begin{bmatrix} y_1 \\ y_2 \\ \cdot \\ \cdot \\ \cdot \\ y_s \end{bmatrix} = \begin{bmatrix} H_{11} & H_{12} & \ldots & H_{1r} \\ H_{21} & H_{22} & \ldots & H_{2r} \\ \cdot & \cdot & & \cdot \\ \cdot & \cdot & & \cdot \\ \cdot & \cdot & & \cdot \\ H_{s1} & H_{s2} & \ldots & H_{sr} \end{bmatrix} \begin{bmatrix} f_1 \\ f_2 \\ \cdot \\ \cdot \\ \cdot \\ f_r \end{bmatrix} \tag{1.25}$$

where

$$y_i = \sum_{j=1}^{r} H_{ij} f_j, \qquad i = 1, 2, \ldots, s \tag{1.26}$$

In abbreviated notation, (1.25) becomes

$$\mathbf{y} = \mathbf{H}\mathbf{f} \tag{1.27}$$

Here $\mathbf{y}$ is an s-vector, $\mathbf{f}$ is an r-vector, and $\mathbf{H}$ is an $s \times r$ operator matrix.

Returning now to the system equation (1.2), we observe that a system may be represented as a nonanticipatory operator if (1) the system is initially at rest, and (2) one is interested only in the output, that is, in the directly observable behavior. If these conditions are satisfied, a system may be viewed simply as a transformation or mapping from the input function space to the output space. In the most general form, such a mapping would be represented by (1.24). Note that when we represent a system as an operator, we implicitly discard the concept of the state of a system.

If the two conditions are not met, it is still possible, though more complicated, to regard a system as an operator (or as an assembly of interconnected operators). This may be accomplished by replacing the initial conditions at t_0 by equivalent inputs applied to the system at t_0^-. If we are interested in some of the state variables, they must be included as additional components in the output vector.

The two approaches to system analysis, that is, with and without the notion of state, have both been effectively used in the solution of important system problems. Although they are able to yield the same results, they may differ markedly in the difficulty attending the solution. Which approach is preferable depends largely on the nature of the problem.

1.4 LINEARITY

One of the most important concepts of system theory is that of linearity. An operator is said to be *linear* if it is both homogeneous and distributive over operands,[4] that is, if for any two vectors $\mathbf{f}$ and $\mathbf{g}$ within the domain of H,

$$H(a\mathbf{f} + b\mathbf{g}) = aH\mathbf{f} + bH\mathbf{g} \tag{1.28}$$

where a and b are arbitrary scalars. A system (or subsystem) is considered linear if its state equations can be expressed in terms of linear operators. If (1.28) is not satisfied, the operator (or any system containing this operator) is said to be *nonlinear*.

Although physical phenomena are never truly linear, a very large number of them can be satisfactorily *approximated* by means of linear models. This is particularly true if the applied forces (inputs) are limited to small variations about a stationary mean.

For a linear system, (1.8) and (1.9) become simply

$$\frac{d}{dt}\mathbf{x}(t) = \mathbf{A}(t)\mathbf{x}(t) + \mathbf{B}(t)\mathbf{u}(t) \tag{1.29}$$

$$\mathbf{y}(t) = \mathbf{C}(t)\mathbf{x}(t) \tag{1.30}$$

where $\mathbf{u}$ is an m-vector, $\mathbf{x}$ is an n-vector, $\mathbf{y}$ is a p-vector, and the $\mathbf{A}$, $\mathbf{B}$, $\mathbf{C}$'s are, respectively, $n \times n$, $n \times m$, and $p \times n$ *linear* operator matrices. With only minor loss of generality, the $\mathbf{A}$, $\mathbf{B}$, $\mathbf{C}$'s may be taken as ordinary matrices.

For linear discrete-time systems, we write instead of (1.29) and (1.30)

$$\mathbf{x}(k + 1) = \mathbf{A}(k)\mathbf{x}(k) + \mathbf{B}(k)\mathbf{u}(k) \tag{1.31}$$

$$\mathbf{y}(k) = \mathbf{C}(k)\mathbf{x}(k) \tag{1.32}$$

The $n \times n$ matrix $\mathbf{A}$ in (1.31) is of particular importance in the system description. This matrix indicates the manner in which the system goes

[4] The term *additive* is used by some authors to describe an operator that is distributive over operands.

through its transitions from state to state in the absence of an input. It will be called the *unit-transition matrix*.

1.5 STATIONARITY

A continuous-time operator H is said to be *stationary* if it obeys the relation

$$\mathbf{y}(t - t_1) = H\{\mathbf{x}(t - t_1)\} \tag{1.33}$$

for all t and t_1, and for all $\mathbf{x}$ within the domain of H. A system is stationary if it can be described by stationary operators. Thus a system that can be represented by a differential equation with constant coefficients is a stationary system. The system of (1.29) and (1.30) is a stationary system if the **A, B, C**'s satisfy (1.33), that is, if the **A, B, C**'s are constant matrices.

A discrete-time operator is stationary if

$$\mathbf{y}(k - i) = H\{\mathbf{x}(k - i)\} \tag{1.34}$$

for all integers k and i, and for all $\mathbf{x}$ in the domain of H. The linear discrete-time system of (1.31) and (1.32) is stationary if its **A, B, C**'s satisfy (1.34).

There is no interrelation between stationarity, linearity, or time discreteness. An operator (or system) may be linear or nonlinear, stationary or nonstationary, discrete-time or continuous-time, in any combination.

1.6 SYSTEM EQUIVALENCE

In a general sense, we regard two systems to be equivalent if they can serve as alternate, equally precise models of a given dynamic phenomenon. It is thus of interest to determine the specific conditions under which two systems can be considered equivalent. We shall distinguish between two kinds of equivalence: *observable equivalence* and *strict equivalence*.

Consider two systems S_1 and S_2 whose input, state, and output vectors are $\mathbf{u}^1$, $\mathbf{x}^1$, $\mathbf{y}^1$, and $\mathbf{u}^2$, $\mathbf{x}^2$, $\mathbf{y}^2$, respectively. The two systems S_1 and S_2 are *observably equivalent* if for all $\mathbf{u}$ in U and all t in Θ, the relation

$$\mathbf{u}^1(t) = \mathbf{u}^2(t) \tag{1.35}$$

implies the relation

$$\mathbf{y}^1(t) = \mathbf{y}^2(t) \tag{1.36}$$

The definition of observable equivalence makes no reference to the state variables of the system; that is, it involves only the system's "exterior." Since it thus cannot account for possible differences in the system's internal behavior, we must introduce an additional, more comprehensive form of equivalence.

In defining the input and output vectors of a system, coordinate systems are selected in the vector spaces U and Y so as to establish a one-to-one correspondence between a "force" and an input coordinate, and between a directly observable behavior characteristic and an output coordinate. Once the "forces" and observable behavior characteristics are identified, the coordinate systems for $\mathbf{u}$ and $\mathbf{y}$ are fixed. In contrast, the coordinate system for the representation of $\mathbf{x}$ in the state space X is not so constrained and may often be selected arbitrarily by the system analyst. Since our interest is in finding equivalent representations for a dynamic phenomenon, any definition of system equivalence involving the state space must be independent of an arbitrarily chosen coordinate system. This suggests the following definition for strict equivalence:

Two systems S_1 and S_2 are *strictly equivalent* if (1) they are observably equivalent, and (2) if, for all $\mathbf{u}$ in U and all t in Θ, the relation

$$\mathbf{u}^1(t) = \mathbf{u}^2(t) \tag{1.37}$$

implies the relation

$$\mathbf{x}^1(t) = \mathbf{F}\mathbf{x}^2(t) \tag{1.38}$$

where $\mathbf{F}$ is a nonsingular, constant $n \times n$ matrix; that is, at any time t the state vectors of two strictly equivalent systems may differ by no more than a constant linear transformation.

1.7 CONTROLLABILITY AND OBSERVABILITY

The state equations

$$\mathbf{y}(t) = G_3(\mathbf{x}(t)) \tag{1.4}$$

$$\mathbf{x}(t) = G_4(\mathbf{x}(t_0); \mathbf{u}_{[t_0, t)}) \tag{1.5}$$

make it possible to determine the system output $\mathbf{y}$ for any time $t > t_0$, given the initial state $\mathbf{x}(t_0)$ and the input $\mathbf{u}$ over $[t_0, t)$. The state vector $\mathbf{x}$ plays an intermediate role in the system characterization; it describes for any time t the current status (state) of the system. In general terms, $\mathbf{y}$ is a function of $\mathbf{x}$, and $\mathbf{x}$ is a function (actually a *functional*) of $\mathbf{u}$ over a specified time interval.

Two questions of particular interest arise in connection with the relationships between $\mathbf{u}$ and $\mathbf{x}$, and between $\mathbf{x}$ and $\mathbf{y}$: (1) whether it is possible to place a given system into any desired state by means of a suitably chosen input over a finite time interval, and (2) whether it is possible to determine the state $\mathbf{x}(t_0)$ from a record of the output over the finite interval $[t_0, t]$. To permit a more precise discussion of these questions, we introduce the following two definitions:[5]

1. A system is said to be *controllable* at time t_0 if it is possible by means of a suitably chosen input to transfer the system from any initial state $\mathbf{x}(t_0)$ in the state space X to any other state in X in a finite interval of time $[t_0, t)$.
2. A system is said to be *observable* at time t_0 if, with the system initially in any state $\mathbf{x}(t_0)$ in the state space X, it is possible to determine this state from observation of the output over a finite interval of time $[t_0, t]$.

For stationary systems, the questions of controllability and observability do not depend on t_0.

For stationary, linear systems the foregoing definitions of controllability and observability can be modified to read as follows:

1. A stationary linear system is *controllable* if it is not possible to transform the system into a strictly equivalent system for which one or more of the state variables (coordinates) x_i are independent of the input $\mathbf{u}$.
2. A stationary linear system is *observable* if it is not possible to transform the system into a strictly equivalent system for which the output $\mathbf{y}$ is independent of one or more of the state variables x_i.

A state variable that is independent of the input is said to be *uncontrollable*. If the output is independent of a state variable, that state variable is said to be *unobservable*.

We note that a state variable that is not controllable (but is observable) may have an effect on the output through its initial value. This appears to contradict our earlier statement that the state variables represent the composite effect of past inputs. We reconcile the two ideas by saying that the state vector represents the composite effect of *all* forces—known and unknown—that have acted on the system in the past. The "unknown" forces, which can, of course, not be part of the input $\mathbf{u}$, are the forces that are either not measured or not measurable by the observer who is formulating the system model. Some such forces are always present in a physical phenomenon to excite any natural modes that cannot be excited by the input $\mathbf{u}$.

On the basis of the definitions of controllability and observability for a linear system it is possible to separate a linear system into four interconnected subsystems (Fig. 1-3). Although the nonobservable subsystems can, by definition, have no direct effect on the output, they may have a possible indirect effect on the output when practical limitations are

[5] Kalman, R., "Mathematical Description of Linear Dynamical Systems," *J. Soc. Indust. Appl. Math., Ser. A: Control*, vol. 1, no. 2, pp. 152–192, 1963.

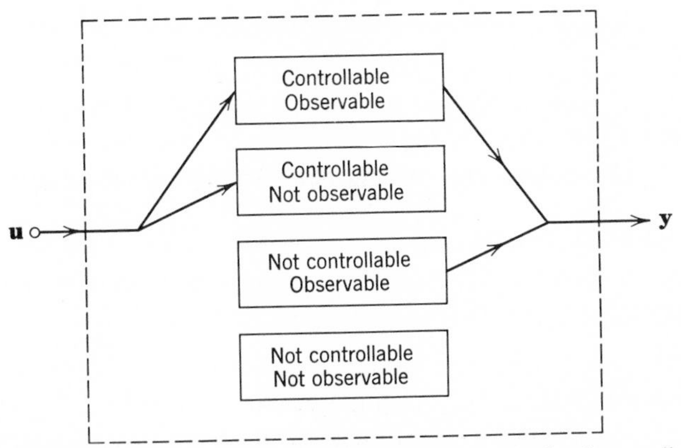

Figure 1-3 Linear system separated into subsystems on the basis of controllability and observability.

considered. Thus, for example, uncontrolled oscillations in a non-observable subsystem could (when referred to the physical dynamic phenomenon represented by the system) cause saturation and destroy the validity of the particular system model. We shall discuss controllability and observability in somewhat more detail in the next chapter.

REFERENCES

Ashby, W. Ross, *An Introduction to Cybernetics*, John Wiley and Sons, New York, 1956.

Brown, R. G. and J. W. Nilsson, *Introduction to Linear System Analysis*, John Wiley and Sons, New York, 1962.

Bunge, Mario, "Causality, Chance, and Law," *Am. Scientist*, vol. 49, no. 4, pp. 432–448, December 1961.

Freeman, H., "A Synthesis Method for Multipole Control Systems," *Trans. AIEE*, vol. 76, part 2, pp. 28–31, March 1957.

Friedman, Bernard, *Principles and Techniques of Applied Mathematics*, John Wiley and Sons, New York, 1956.

Gilbert, E. G., "Controllability and Observability in Multivariable Control Systems," *J. Soc. Ind. Appl. Math., Ser. A: Control*, vol. 1, no. 2, pp. 128–151, 1963.

Kalman, R., "Mathematical Description of Linear Dynamical Systems," *J. Soc. Ind. Appl. Math., Ser. A: Control*, vol. 1, no. 2, pp. 152–192, 1963.

Kalman, R., "On the General Theory of Control Systems," *Proc. 1st Intern. Congr. Autom. Control, Moscow*, 1960, vol. 1, Butterworths, London, pp. 481–492, 1961.

Kalman, R., Y. Ho, and K. Narendra, "Controllability of Linear Dynamical Systems," *Contrib. Differential Equations*, vol. 1, no. 2, pp. 189–213, 1963.

Zadeh, L. A., "From Circuit Theory to System Theory," *Proc. IRE*, vol. 50, no. 5, pp. 856–865, May 1962.

Zadeh, L. A. and C. A. Desoer, *Linear System Theory*, McGraw-Hill Book Co., New York, 1963.

CHAPTER 2

Analysis in Discrete-Time Domain

2.1 INTRODUCTION

The analysis of linear discrete-time systems can be carried out in either of two different ways. One of these is based on techniques that are directly applicable in the time domain; the other employs a set of techniques in a *transformed* domain. Although the time-domain techniques are more direct than those of the transformed domain, the latter may involve simpler computations for certain classes of problems. Both approaches will be examined. This chapter is concerned with time-domain techniques; techniques in the transformed domain are described in Chapter 3. In both chapters only discrete-time systems that are deterministic and linear are considered.

2.2 SYSTEM WEIGHTING SEQUENCE

It is important at the outset to clarify our notation of time. As indicated in the linear discrete-time state equations (1.31) and (1.32), an input $\mathbf{u}(k)$ can affect neither the state $\mathbf{x}(k)$ nor the output $\mathbf{y}(k)$. The first instant at which the effect of $\mathbf{u}(k)$ appears in the state and the output of the system is at time $k + 1$. In effect, we may say that $\mathbf{y}(k)$ *follows* $\mathbf{u}(k - 1)$ *in time but precedes* $\mathbf{u}(k)$. This time convention is consistent with our definition of a nonanticipatory system given in Section 1.1. We shall employ it in all the derivations to follow.

We consider a stationary, linear, deterministic, discrete-time system that is fully relaxed at $k < 0$. The system is assumed to possess only a single input and a single output; that is, both $\mathbf{u}$ and $\mathbf{y}$ are 1-vectors. At time $k = 0$, we apply an input $u(0) = 1$ and observe the output y over the interval $1 \leq k < \infty$. Let the output be

$$y(k) = h(k), \qquad k = 1, 2, \ldots, \infty \tag{2.1}$$

where the $h(k)$ are real numbers. Since the system is linear by definition, we may alternately apply an input $u(0)$ *of any magnitude* at time 0 and obtain an output at time k:

$$y(k) = h(k)u(0), \qquad k = 1, 2, \ldots, \infty \tag{2.2}$$

Now suppose that instead of applying an input $u(0)$ at time 0, we apply an input $u(1)$ at time 1. Since in addition to being linear the system is also stationary, we now obtain at time k an output

$$y(k) = h(k-1)u(1) \tag{2.3}$$

where now $k = 2, 3, \ldots, \infty$. More generally, a single input $u(i)$ applied to a relaxed system at an arbitrary instant i within the discrete-time domain Θ will thus yield an output

$$y(k) = h(k-i)u(i) \tag{2.4}$$

for $k = i + 1,\ i + 2, \ldots,\ \infty$. For a nonanticipatory system (and in accordance with our time convention) it is necessary that

$$h(k) = 0 \quad \text{for all } k < 1 \tag{2.5}$$

Since the system is linear, a sequence of inputs $\{u(i)\}$, where $0 \leq i < \infty$, will result in a composite system output at time k that is a linear summation of the weighted contributions from each $u(i)$ for $i < k$. Thus

$$\boxed{y(k) = \sum_{i=0}^{k-1} h(k-i)u(i), \qquad k = 1, 2, \ldots, \infty} \tag{2.6}$$

Equation (2.6) is known as the *convolution summation* of the system. It is the discrete-time counterpart of the convolution integral for continuous-time systems.

If we let $i = k - j$ in (2.6), we obtain the alternate form,

$$\boxed{y(k) = \sum_{j=1}^{k} h(j)u(k-j)} \tag{2.7}$$

We observe that the $h(k - i)$ in (2.6) determine the weights with which the input values $u(i)$, $i = 0, 1, \ldots, k - 1$, contribute to the output $y(k)$ at time k. For this reason the sequence $\{h(k)\}$, $(k = 1, 2, \ldots, \infty)$ is known as the *weighting sequence* of the system. Given a deterministic, linear, stationary, discrete-time system that is initially relaxed, the output $y(k)$ resulting from any input $\{u(i)\}$, $i = 0, 1, \ldots, k - 1$, can be found at once by means of (2.6) or (2.7).

If a system is nonstationary, (2.6) takes on the form

$$y(k) = \sum_{i=0}^{k-1} h(k, k - i)u(i) \tag{2.8}$$

where $h(k, k - i)$ is the weight assigned at time k to an input applied at time i, that is, $k - i$ seconds earlier. It is important to note that for a nonstationary system the weighting sequence $\{h(k, j)\}$, as employed in (2.8), is a function of two arguments, the first, k, indicating "present time" and the second, j, indicating the "age" of the input to be weighted.

Example

Consider a system characterized by the (nonstationary) weighting sequence

$$h(k, j) = 15(3)^{-k}\, 2^{j}$$

To be able to utilize this weighting sequence in an expression such as (2.8), we must first replace j by $k - i$. We obtain

$$h(k, k - i) = 15(3)^{-k}\, 2^{k-i}$$

and can then write

$$y(k) = \sum_{i=0}^{k-1} 15(3)^{-k}\, 2^{k-i} u(i)$$

Equations (2.6)–(2.8) apply only to initially relaxed systems for which both **u** and **y** are one-dimensional. In the general case where **u** and **y** are of dimensions m and p, respectively, we write

$$y_i(k) = \sum_{r=0}^{k-1} \sum_{j=1}^{m} h_{ij}(k - r)u_j(r), \qquad i = 1, 2, \ldots, p \tag{2.9}$$

if the system is stationary, and

$$y_i(k) = \sum_{r=0}^{k-1} \sum_{j=1}^{m} h_{ij}(k, k - r)u_j(r) \tag{2.10}$$

if the system is nonstationary.

In matrix form, (2.9) and (2.10) become, respectively,

$$\mathbf{y}(k) = \sum_{r=0}^{k-1} \mathbf{H}(k - r)\mathbf{u}(r) \tag{2.11}$$

$$\mathbf{y}(k) = \sum_{r=0}^{k-1} \mathbf{H}(k, k - r)\mathbf{u}(r) \tag{2.12}$$

The matrix **H** is called the *weighting sequence matrix* of the system.

2.3 TRANSMISSION MATRICES

It is instructive to write (2.6) in matrix form as follows:

$$\begin{bmatrix} y(1) \\ y(2) \\ y(3) \\ \cdot \\ \cdot \\ \cdot \\ y(k) \end{bmatrix} = \begin{bmatrix} h(1) & 0 & 0 & 0 & \dots & 0 \\ h(2) & h(1) & 0 & 0 & \dots & 0 \\ h(3) & h(2) & h(1) & 0 & \dots & 0 \\ \cdot & \cdot & \cdot & \cdot & & \cdot \\ \cdot & \cdot & \cdot & \cdot & & \cdot \\ \cdot & \cdot & \cdot & \cdot & & \cdot \\ h(k) & \cdot & \cdot & \cdot & \cdot & h(1) \end{bmatrix} \begin{bmatrix} u(0) \\ u(1) \\ u(2) \\ \cdot \\ \cdot \\ \cdot \\ u(k-1) \end{bmatrix} \tag{2.13}$$

The triangular $k \times k$ matrix is known as the *transmission matrix* of the system. It has the important properties

$$h_{ij} = 0 \qquad \text{for } i < j \tag{2.14}$$

$$h_{ij} = h_{i+1,j+1} \quad \text{for all } k \geq i \geq j \geq 1 \tag{2.15}$$

It follows from (2.15) that h_{ij} is a function of only the difference $i - j$; that is, we can write

$$h_{ij} = h(i - j + 1) \tag{2.16}$$

Consider now two cascaded systems (both initially relaxed) described by the stationary transmission matrices $[h_{ij}]$ and $[g_{ij}]$, as shown in Fig. 2-1, such that the output of the first serves as the input of the second.

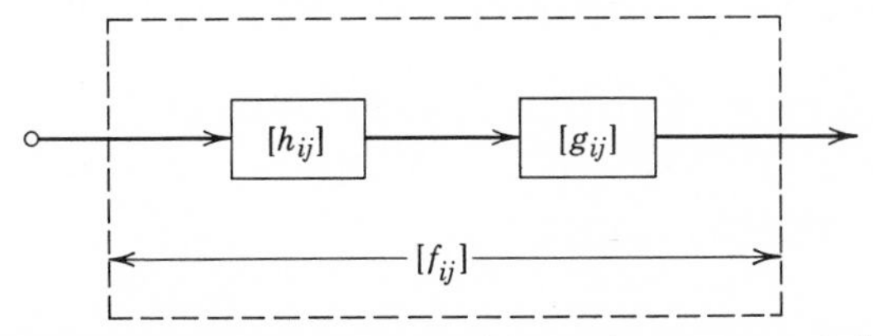

Figure 2-1 Two cascaded systems described by stationary transmission matrices.

From (2.11) the transmission matrix $[f_{ij}]$ for the combined systems must be given by the matrix product:

$$[f_{ij}] = [g_{ij}][h_{ij}] \tag{2.17}$$

that is,

$$f_{ij} = \sum_{s=1}^{k} g_{is} h_{sj} \tag{2.18}$$

subject to the conditions (2.14) and (2.15).

If the order in which the two systems are cascaded is interchanged, then

$$f_{ij} = \sum_{s=1}^{k} h_{is} g_{sj} \tag{2.19}$$

From (2.14) and (2.18),

$$f_{ij} = g_{ij} h_{jj} + g_{i,j+1} h_{j+1,j} + \cdots + g_{i,i-1} h_{i-1,j} + g_{ii} h_{ij} \tag{2.20}$$

Using (2.16), we obtain

$$f_{ij} = g(i-j+1)h(1) + g(i-j)h(2) + g(i-j-1)h(3) + \cdots + g(2)h(i-j) + g(1)h(i-j+1) \tag{2.21}$$

Expanding (2.19) in similar manner yields

$$f_{ij} = h_{ij} g_{jj} + h_{i,j+1} g_{j+1,j} + \cdots + h_{i,i-1} g_{i-1,j} + h_{ii} g_{ij} \tag{2.22}$$

$$f_{ij} = h(i-j+1)g(1) + h(i-j)g(2) + \cdots + h(2)g(i-j) + h(1)g(i-j+1) \tag{2.23}$$

Comparison of (2.23) and (2.21) shows them to be identical. We conclude that matrices satisfying (2.14) and (2.15) are commutative. This implies in turn that the cascading of deterministic, linear, stationary, discrete-time systems is independent of the order in which the cascading is arranged.

For the nonstationary case, (2.8) yields the transmission-matrix equation:

$$\begin{bmatrix} y(1) \\ y(2) \\ y(3) \\ \cdot \\ \cdot \\ \cdot \\ y(k) \end{bmatrix} = \begin{bmatrix} h(1,1) & 0 & 0 & \cdots & 0 \\ h(2,2) & h(2,1) & 0 & \cdots & 0 \\ h(3,3) & h(3,2) & h(3,1) & \cdots & 0 \\ \cdot & \cdot & \cdot & & \cdot \\ \cdot & \cdot & \cdot & & \cdot \\ \cdot & \cdot & \cdot & & \cdot \\ h(k,k) & h(k,k-1) & h(k,k-2) & \cdots & h(k,1) \end{bmatrix} \begin{bmatrix} u(0) \\ u(1) \\ u(2) \\ \cdot \\ \cdot \\ \cdot \\ u(k-1) \end{bmatrix} \tag{2.24}$$

Since (2.15) does not hold for (2.24), nonstationary transmission matrices are not commutative and hence neither are nonstationary systems.

There is no convenient way for writing a transmission matrix to correspond to the general (m, p) cases (2.9) and (2.10).

The transmission matrix of the two systems in parallel (Fig. 2-2), for which $u = u^1 = u^2$ and $y = y^1 + y^2$, is simply the sum of their respective separate transmission matrices; that is,

$$f_{ij} = h_{ij} + g_{ij} \tag{2.25}$$

for all i and j, $k \geq i \geq j \geq 1$.

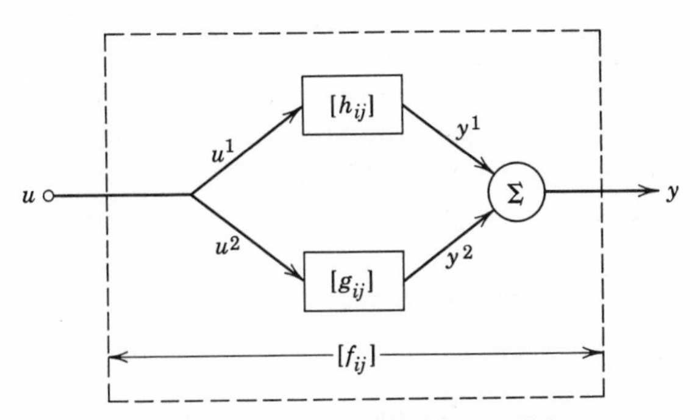

Figure 2-2 Two systems in parallel.

2.4 CHARACTERIZATION IN TERMS OF STATE VARIABLES

If a system to be analyzed is not initially relaxed, the state variable approach is generally preferred over the weighting function approach. From (1.31) and (1.32), a deterministic, linear, stationary, discrete-time system can be characterized by the two state equations:

$$\mathbf{x}(k + 1) = \mathbf{Ax}(k) + \mathbf{Bu}(k) \tag{2.26}$$

$$\mathbf{y}(k) = \mathbf{Cx}(k) \tag{2.27}$$

We shall now derive an expression for $\mathbf{x}(k)$ at any $k > 0$, given $\mathbf{x}(0)$ and $\mathbf{u}(k)$ over the interval $[0, k - 1]$. Substituting successively $k = 0, 1$, and 2 in (2.26), we obtain

$$\begin{aligned}
\mathbf{x}(1) &= \mathbf{Ax}(0) + \mathbf{Bu}(0) \\
\mathbf{x}(2) &= \mathbf{Ax}(1) + \mathbf{Bu}(1) \\
&= \mathbf{A}^2\mathbf{x}(0) + \mathbf{ABu}(0) + \mathbf{Bu}(1) \\
\mathbf{x}(3) &= \mathbf{Ax}(2) + \mathbf{Bu}(2) \\
&= \mathbf{A}^3\mathbf{x}(0) + \mathbf{A}^2\mathbf{Bu}(0) + \mathbf{ABu}(1) + \mathbf{Bu}(2)
\end{aligned}$$

Hence for any $k > 0$,

$$\mathbf{x}(k) = \mathbf{A}^k\mathbf{x}(0) + \sum_{j=0}^{k-1} \mathbf{A}^j\mathbf{Bu}(k - j - 1) \tag{2.28}$$

where $\mathbf{A}^k$ is the k-fold matrix product $\mathbf{A} \times \mathbf{A} \times \cdots$. We shall call $\mathbf{A}^k$ the *fundamental matrix*[1] of the system and denote it by $\mathbf{\Phi}(k)$. Then

$$\boxed{\mathbf{x}(k) = \mathbf{\Phi}(k)\mathbf{x}(0) + \sum_{j=0}^{k-1} \mathbf{\Phi}(j)\mathbf{B}\mathbf{u}(k-j-1)} \tag{2.29}$$

We refer to (2.29) as the *system transition equation.* It gives the state vector at time $k > 0$ as the sum of two major terms, one representing the contribution due to the initial state $\mathbf{x}(0)$, and the other, the contribution due to the input $\mathbf{u}$ over the interval $[0, k-1]$.

For the case of *zero initial conditions*, we set $\mathbf{x}(0) = 0$ in (2.29) and write for (2.27)

$$\mathbf{y}(k) = \sum_{j=0}^{k-1} \mathbf{C}\mathbf{\Phi}(j)\mathbf{B}\mathbf{u}(k-j-1), \qquad k = 1, 2, \ldots \tag{2.30}$$

Let $j = k - r - 1$. Then

$$\mathbf{y}(k) = \sum_{r=0}^{k-1} \mathbf{C}\mathbf{\Phi}(k-r-1)\mathbf{B}\mathbf{u}(r) \tag{2.31}$$

Comparison with $\mathbf{y}(k)$ in (2.11) shows that

$$\mathbf{H}(k-r) = \mathbf{C}\mathbf{\Phi}(k-r-1)\mathbf{B} \tag{2.32}$$

$$\boxed{\mathbf{H}(k-r) = \mathbf{C}\mathbf{A}^{k-r-1}\mathbf{B}, \qquad k \geq r+1} \tag{2.33}$$

The last equation gives the weighting sequence matrix $\mathbf{H}$ in terms of the state equation matrices $\mathbf{A}$, $\mathbf{B}$, and $\mathbf{C}$.

A system transition equation can also be derived for the nonstationary case. From (1.31),

$$\begin{aligned}
\mathbf{x}(1) &= \mathbf{A}(0)\mathbf{x}(0) + \mathbf{B}(0)\mathbf{u}(0) \\
\mathbf{x}(2) &= \mathbf{A}(1)\mathbf{x}(1) + \mathbf{B}(1)\mathbf{u}(1) \\
&= \mathbf{A}(1)\mathbf{A}(0)\mathbf{x}(0) + \mathbf{A}(1)\mathbf{B}(0)\mathbf{u}(0) + \mathbf{B}(1)\mathbf{u}(1) \\
\mathbf{x}(3) &= \mathbf{A}(2)\mathbf{x}(2) + \mathbf{B}(2)\mathbf{u}(2) \\
&= \mathbf{A}(2)\mathbf{A}(1)\mathbf{A}(0)\mathbf{x}(0) + \mathbf{A}(2)\mathbf{A}(1)\mathbf{B}(0)\mathbf{u}(0) \\
&\qquad + \mathbf{A}(2)\mathbf{B}(1)\mathbf{u}(1) + \mathbf{B}(2)\mathbf{u}(2) \\
&\text{etc.}
\end{aligned} \tag{2.34}$$

If for $k > j$ we define

$$\mathbf{\Phi}(k, j) = \prod_{i=j}^{k-1} \mathbf{A}(i) = \mathbf{A}(k-1)\mathbf{A}(k-2)\cdots\mathbf{A}(j+1)\mathbf{A}(j) \tag{2.35}$$

and

$$\mathbf{\Phi}(k, k) = \mathbf{I} \quad \text{(identity matrix)} \tag{2.36}$$

[1] The reader is cautioned to note that some writers, instead of using the term "fundamental matrix," use the term "transition matrix" to describe the matrix $\mathbf{\Phi}(k)$ rather than the matrix $\mathbf{A}$.

we can write

$$\boxed{\mathbf{x}(k) = \mathbf{\Phi}(k, 0)\mathbf{x}(0) + \sum_{j=0}^{k-1} \mathbf{\Phi}(k, j + 1)\mathbf{B}(j)\mathbf{u}(j)} \tag{2.37}$$

Note that the matrices in (2.35) are in general not commutable.

The first term on the right side of (2.37) is the contribution from the initial state $\mathbf{x}(0)$, and the second term is the contribution from the input over the interval $[0, k - 1]$.

If $\mathbf{x}(0) = 0$, we can write for the system output, using (1.32),

$$\mathbf{y}(k) = \sum_{j=0}^{k-1} \mathbf{C}(k)\mathbf{\Phi}(k, j + 1)\mathbf{B}(j)\mathbf{u}(j) \tag{2.38}$$

for all integers $k > 0$.

If we now compare (2.12) with (2.38), we find

$$\boxed{\mathbf{H}(k, k - j) = \mathbf{C}(k)\mathbf{\Phi}(k, j + 1)\mathbf{B}(j)} \tag{2.39}$$

From (2.35), for $k \geq l \geq j$,

$$\prod_{i=j}^{k-1} \mathbf{A}(i) = \prod_{i=l}^{k-1} \mathbf{A}(i) \prod_{q=j}^{l-1} \mathbf{A}(q) \tag{2.40}$$

Hence

$$\boxed{\mathbf{\Phi}(k, j) = \mathbf{\Phi}(k, l)\mathbf{\Phi}(l, j)} \tag{2.41}$$

If we let $j = k$ in (2.41),

$$\mathbf{\Phi}(k, k) = \mathbf{\Phi}(k, l)\mathbf{\Phi}(l, k) \tag{2.42}$$

Using (2.36), we have

$$\mathbf{\Phi}(k, l)\mathbf{\Phi}(l, k) = \mathbf{I}$$

If for $k < j$ we now define

$$\mathbf{\Phi}(k, j) = \mathbf{A}^{-1}(k)\mathbf{A}^{-1}(k + 1) \ldots \mathbf{A}^{-1}(j - 1) \tag{2.43}$$

then, *provided these inverses of the* $\mathbf{A}(k)$ *matrix exist,*

$$\boxed{\mathbf{\Phi}(l, k) = \mathbf{\Phi}^{-1}(k, l)} \tag{2.44}$$

From (2.44) follows the interesting result that the weighting sequence matrix (2.39) can be written as the product of two matrices such that

$$\boxed{\mathbf{H}(k, k - j) = \mathbf{H}_1(k)\mathbf{H}_2(j)} \tag{2.45}$$

where

$$\mathbf{H}_1(k) = \mathbf{C}(k)\mathbf{\Phi}(k, 0) \tag{2.46a}$$

$$\mathbf{H}_2(j) = \mathbf{\Phi}(0, j + 1)\mathbf{B}(j) \tag{2.46b}$$

2.5 FORMULATION OF STATE EQUATIONS

A common means for describing a linear, discrete-time system is in terms of an nth-order difference equation. We shall now give a method for obtaining the state equations for such a system, given its nth-order difference equation.

For a single-input, single-output, linear, deterministic system ($m = p = 1$), the standard form for the nth-order difference equation is

$$a_0y(k) + a_1y(k-1) + \cdots + a_ry(k-r) = b_1u(k-1) + b_2u(k-2) + \cdots + b_su(k-s), \qquad k = 1, 2, \ldots \tag{2.47}$$

where r and s are positive integers. Note that this way of writing the equation is consistent with our previously adopted time labeling convention according to which $y(k)$ follows $u(k-1)$ but precedes $u(k)$.

Before proceeding further it is desirable to rewrite (2.47) in the following form,

$$a_0y(k) + a_1y(k-1) + \cdots + a_ny(k-n) = b_1u(k-1) + \cdots + b_su(k-s) \tag{2.48}$$

in which

$$n = \max(r, s) \tag{2.49}$$

and $a_i = 0$ for any $i > r$. Equation (2.48) can also be written in the alternate form,

$$(a_0 + a_1E^{-1} + \cdots + a_nE^{-n})y(k) = (b_1E^{-1} + \cdots + b_sE^{-s})u(k) \tag{2.50}$$

where E is the linear advance operator defined by

$$E^{\nu}f(k) = f(k+\nu) \tag{2.51}$$

for any integer ν.

The system described by (2.48) can be represented in state-equation form in a variety of equivalent ways. Since we have an nth-order difference equation, n initial conditions are required together with the input $u_{[0,k)}$ to specify uniquely the output $y(k)$ for all $k > 0$. It is this value of n which determines the dimensionality of the state vector $\mathbf{x}$.

We introduce a new variable $v(k)$ by writing in place of (2.50),

$$(a_0 + a_1E^{-1} + \cdots + a_nE^{-n})v(k) = u(k) \tag{2.52}$$

and

$$(b_1E^{-1} + \cdots + b_sE^{-s})v(k) = y(k) \tag{2.53}$$

From (2.52),

$$v(k) = \frac{1}{a_0}[u(k) - a_nv(k-n) - a_{n-1}v(k-n+1) - \cdots - a_1v(k-1)] \tag{2.54}$$

Now let

$$
\begin{aligned}
x_1(k) &= v(k-n) \\
x_2(k) &= v(k-n+1) \\
x_3(k) &= v(k-n+2) \\
&\;\vdots \\
x_n(k) &= v(k-1)
\end{aligned}
\tag{2.55}
$$

We can then write

$$
\begin{aligned}
x_1(k+1) &= x_2(k) \\
x_2(k+1) &= x_3(k) \\
&\;\vdots \\
x_{n-1}(k+1) &= x_n(k) \\
x_n(k+1) &= v(k) = \frac{1}{a_0}[u(k) - a_n x_1(k) - a_{n-1}x_2(k) - \cdots - a_1 x_n(k)]
\end{aligned}
\tag{2.56}
$$

In matrix form,

$$\mathbf{x}(k+1) = \mathbf{A}\mathbf{x}(k) + \mathbf{B}\mathbf{u}(k) \tag{2.57}$$

where

$$
\mathbf{A} = \begin{bmatrix}
0 & 1 & 0 & \cdots & 0 & 0 \\
0 & 0 & 1 & \cdots & 0 & 0 \\
\vdots & & & & & \vdots \\
0 & 0 & 0 & \cdots & 0 & 1 \\
-\frac{a_n}{a_0} & -\frac{a_{n-1}}{a_0} & -\frac{a_{n-2}}{a_0} & \cdots & -\frac{a_2}{a_0} & -\frac{a_1}{a_0}
\end{bmatrix}
\tag{2.58}
$$

$$
\mathbf{B} = \begin{bmatrix}
0 \\ 0 \\ \vdots \\ \frac{1}{a_0}
\end{bmatrix}
\tag{2.59}
$$

and $\mathbf{u}(k)$ is a 1×1 matrix.

Returning to (2.53) we write

$$y(k) = b_1 v(k-1) + b_2 v(k-2) + \cdots + b_s v(k-s) \tag{2.60}$$

Substituting from (2.55), we have

$$y(k) = b_1 x_n(k) + b_2 x_{n-1}(k) + \cdots + b_s x_{n+1-s} \tag{2.61}$$

In matrix form

$$\mathbf{y}(k) = \mathbf{C}\mathbf{x}(k) \tag{2.62}$$

where **C** is a row matrix,

$$\mathbf{C} = [b_n \ldots b_{i+1} b_i \ldots b_1] \tag{2.63}$$

in which

$$b_i = 0 \quad \text{for } i > s \tag{2.64}$$

Equations (2.58), (2.59), and (2.63) give the desired state equation matrices **A**, **B**, and **C** for the system of (2.47). The output for any $k > 0$ and

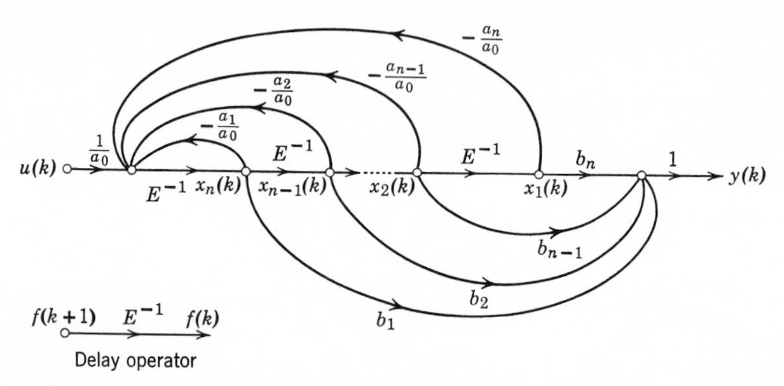

Figure 2-3 Signal flow graph representation showing state variables of single-input, single-output system.

any given set of initial conditions $\mathbf{x}(0)$ can be uniquely calculated with the aid of (2.27) and (2.29). Note that the a_i and b_i may be either constant or may vary with k; that is, the equations apply to both the stationary and the nonstationary case. A signal flow graph representation of this system is given in Fig. 2-3. We shall refer to a diagram of this type also as a *state-transition diagram.*

As previously indicated, the state equation matrices (2.58), (2.59), and (2.63) for the system of (2.47) are not unique. To illustrate this, we shall now derive another, equivalent set of matrices for this system.

By a slight rearrangement of (2.48), we can write

$$y(k) = \frac{b_1}{a_0} u(k-1) + \frac{b_2}{a_0} u(k-2) + \cdots + \frac{b_s}{a_0} u(k-s) - \frac{a_1}{a_0} y(k-1) - \frac{a_2}{a_0} y(k-2) - \cdots - \frac{a_n}{a_0} y(k-n) \tag{2.65}$$

To introduce the state variables we let $y(k) = x_1(k)$ and write

$$x_1(k) = y(k) = \frac{b_1}{a_0} u(k-1) - \frac{a_1}{a_0} x_1(k-1) + x_2(k-1) \tag{2.66}$$

where

$$\begin{aligned} x_2(k-1) = {} & \frac{b_2}{a_0} u(k-2) + \frac{b_3}{a_0} u(k-3) + \cdots + \frac{b_s}{a_0} u(k-s) \\ & - \frac{a_2}{a_0} x_1(k-2) - \frac{a_3}{a_0} x_1(k-3) - \cdots - \frac{a_n}{a_0} x_1(k-n) \end{aligned} \tag{2.67}$$

In this manner we obtain, starting with (2.66) and (2.67);

$$\begin{aligned} x_1(k+1) &= \frac{b_1}{a_0} u(k) - \frac{a_1}{a_0} x_1(k) + x_2(k) \\ x_2(k+1) &= \frac{b_2}{a_0} u(k) - \frac{a_2}{a_0} x_1(k) + x_3(k) \\ &\;\;\vdots \\ x_{n-1}(k+1) &= \frac{b_{n-1}}{a_0} u(k) - \frac{a_{n-1}}{a_0} x_1(k) + x_n(k) \\ x_n(k+1) &= \frac{b_n}{a_0} u(k) - \frac{a_n}{a_0} x_1(k) \end{aligned} \tag{2.68}$$

From (2.68) and (2.66) we can at once write the state equations (2.26) and (2.27). We find

$$\mathbf{A} = \begin{bmatrix} -\frac{a_1}{a_0} & 1 & 0 & \cdots & 0 \\ -\frac{a_2}{a_0} & 0 & 1 & \cdots & 0 \\ \vdots & \vdots & \vdots & & \vdots \\ -\frac{a_{n-1}}{a_0} & 0 & 0 & \cdots & 1 \\ -\frac{a_n}{a_0} & 0 & 0 & \cdots & 0 \end{bmatrix} \tag{2.69}$$

$$\mathbf{B} = \begin{bmatrix} \dfrac{b_1}{a_0} \\ \dfrac{b_2}{a_0} \\ \cdot \\ \cdot \\ \cdot \\ \dfrac{b_n}{a_0} \end{bmatrix} \tag{2.70}$$

$$\mathbf{C} = [1 \quad 0 \quad 0 \quad \dots \quad 0] \tag{2.71}$$

The state transition diagram based on the use of this alternate set of state variables is shown in Fig. 2-4.

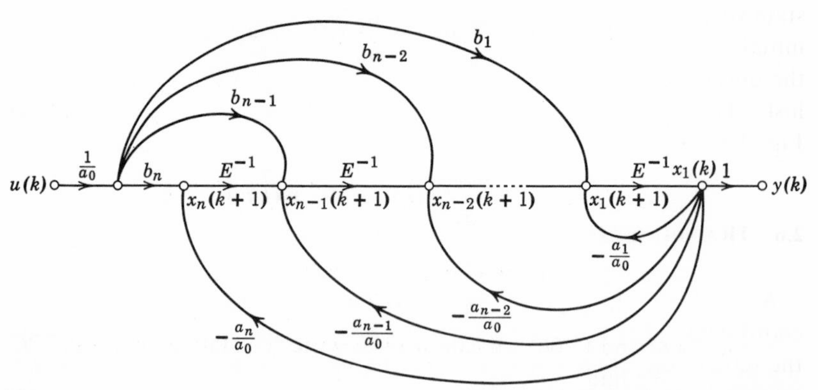

Figure 2-4 System of Fig. 2-3 when represented in terms of an alternate set of state variables.

The system of Fig. 2-4 is strictly equivalent to that of Fig. 2-3. Indeed, any two systems are strictly equivalent if their state vectors at any time k differ by no more than a nonsingular constant linear transformation. This is true for stationary as well as nonstationary systems.

In both Fig. 2-3 and Fig. 2-4 it is noted that each state variable appears as the output of a distinct unit-time delay element. For an nth-order system, n such delay elements are required. Quite clearly, these delay elements are the "memory cells" that store the n coordinate numbers of the state vector $\mathbf{x}$ from one instant of time to the next. If the representation of an nth-order system utilizes more than n delay elements, it is always possible to eliminate those in excess of n.

A system described by a difference equation such as (2.47) is of order n, where

$$n = \max(r, s) \tag{2.49}$$

If all the b_i in (2.47) are zero, the system is "unforced"; that is, the output is solely a function of the initial conditions and independent of any input. This can be seen clearly by referring to Fig. 2-4. We also note from Fig. 2-3 that the representation (2.58), (2.59), and (2.63) is now *not valid* since with all $b_i = 0$, the output—as given by (2.62) and (2.63)—would always appear to be zero (a fallacy). Hence this particular representation cannot be used if all the $b_i = 0$.

In (2.47), all the a_i may be zero except for a_0. In every case in which $r < s$, the determinant of $\mathbf{A}(k)$ will vanish. Hence $\mathbf{A}(k)$ does not have an inverse, and $\mathbf{\Phi}(k, j)$ is, therefore, defined only for $k \geq j$. Specifically this implies that it is not possible to go backward in time; that is, given a state $\mathbf{x}(k)$ plus knowledge of the input $\mathbf{u}_{[0,k)}$, we are *not* able to determine the state $\mathbf{x}(0)$ for $k > 0$. The reason for this is that when $r < s$, only part of the initial-state information is retained into the future. The other part affects the output for only a finite time into the future and then is irretrievably lost. This characteristic of a system can be seen readily by examining Fig. 2-3; it is, however, obscured in Fig. 2-4.

2.6 TRANSFORMATION OF COORDINATES

A set of very simple relations governs the transformation of state-space coordinates. Let $\mathbf{x}^1$ be the state in the given coordinate system and $\mathbf{x}^2$ be the state in the transformed coordinate system. Denoting the nonsingular transformation by $\mathbf{F}$, we have

$$\mathbf{x}^2 = \mathbf{F}\mathbf{x}^1 \tag{2.72}$$

Also,

$$\mathbf{x}^1(k+1) = \mathbf{A}_1(k)\mathbf{x}^1(k) + \mathbf{B}_1(k)\mathbf{u}(k) \tag{2.73a}$$

$$\mathbf{y}(k) = \mathbf{C}_1(k)\mathbf{x}^1(k) \tag{2.73b}$$

and

$$\mathbf{x}^2(k+1) = \mathbf{A}_2(k)\mathbf{x}^2(k) + \mathbf{B}_2(k)\mathbf{u}(k) \tag{2.74a}$$

$$\mathbf{y}(k) = \mathbf{C}_2(k)\mathbf{x}^2(k) \tag{2.74b}$$

Since $\mathbf{F}$ is nonsingular, we can write

$$\mathbf{x}^1 = \mathbf{F}^{-1}\mathbf{x}^2 \tag{2.75}$$

Substituting in (2.73) yields

$$\mathbf{F}^{-1}\mathbf{x}^2(k+1) = \mathbf{A}_1(k)\mathbf{F}^{-1}\mathbf{x}^2(k) + \mathbf{B}_1(k)\mathbf{u}(k) \tag{2.76a}$$

$$\mathbf{y}(k) = \mathbf{C}_1(k)\mathbf{F}^{-1}\mathbf{x}^2(k) \tag{2.76b}$$

Premultiplying (2.76a) by $\mathbf{F}$ and comparing (2.76) with (2.74) then shows that

$$\mathbf{A}_2(k) = \mathbf{F}\mathbf{A}_1(k)\mathbf{F}^{-1} \tag{2.77}$$

$$\mathbf{B}_2(k) = \mathbf{F}\mathbf{B}_1(k) \tag{2.78}$$

$$\mathbf{C}_2(k) = \mathbf{C}_1(k)\mathbf{F}^{-1} \tag{2.79}$$

From (2.35) it follows that

$$\boldsymbol{\Phi}_2(k, j) = \mathbf{F}\boldsymbol{\Phi}_1(k, j)\mathbf{F}^{-1} \tag{2.80}$$

There is one transformation that is of particular interest. A theorem in matrix algebra[2] tells us that if a $n \times n$ matrix $\mathbf{A}$ has distinct eigenvalues, $\lambda_1, \lambda_2, \ldots, \lambda_n$, there exists a transformation $\mathbf{Q}$ such that

$$\mathbf{Q}^{-1}\mathbf{A}\mathbf{Q} = \begin{bmatrix} \lambda_1 & & & & 0 \\ & \lambda_2 & & & \\ & & \cdot & & \\ & & & \cdot & \\ & & & & \cdot \\ 0 & & & & \lambda_n \end{bmatrix} \tag{2.81}$$

The eigenvalues are the solutions of the nth-order *characteristic equation*

$$|\mathbf{A} - \lambda\mathbf{I}| = 0 \tag{2.82}$$

where $\mathbf{I}$ is the identity matrix. The columns of $\mathbf{Q}$ are the eigenvectors of $\mathbf{A}$, determined up to a scalar multiple. Therefore, a system (2.73) whose unit transition matrix $\mathbf{A}$ has distinct eigenvalues can always be represented in the so-called *normal form*

$$\mathbf{x}^2(k + 1) = \mathcal{A}(k)\mathbf{x}^2(k) + \mathcal{B}(k)\mathbf{u}(k) \tag{2.83a}$$

$$\mathbf{y}(k) = \mathcal{C}(k)\mathbf{x}^2(k), \tag{2.83b}$$

where the $\mathcal{A}(k)$, $\mathcal{B}(k)$, and $\mathcal{C}(k)$ result from (2.77–2.79) when the transformation $\mathbf{Q}^{-1}$ is used for $\mathbf{F}$.

If the eigenvectors of $\mathbf{A}$ are scaled so that they have unit Euclidean length, that is, if

$$\|x\| = \sqrt{x_1^2 + x_2^2 + \cdots + x_n^2} = 1 \tag{2.84}$$

the eigenvectors are said to be *normalized.* The normal form (2.83) may be made *unique* by using normalized eigenvectors for the columns of $\mathbf{Q}$ and

[2] R. Bellman, *Introduction to Matrix Analysis*, McGraw-Hill Book Co., 1960.

by arranging the eigenvalues of A in increasing order of magnitude. Eigenvalues of like magnitude may be arranged in order of increasing angle.

Example

Consider the system described by the difference equation:

$$y(k) = \tfrac{5}{2}y(k-1) - \tfrac{3}{2}y(k-2) + 7(2)^{4-k}u(k-1) - 19(2)^{4-k}u(k-2)$$

This equation is similar in form to (2.47). Since $r = s = 2$, we need only two state variables to represent the system.

Let $x_1(k) = y(k)$. Then

$$y(k) = x_1(k) = \tfrac{5}{2}x_1(k-1) + 7(2)^{4-k}u(k-1) + x_2(k-1)$$
$$x_2(k-1) = -\tfrac{3}{2}x_1(k-2) - 19(2)^{4-k}u(k-2)$$

By shifting to $k+1$,

$$x_1(k+1) = \tfrac{5}{2}x_1(k) + x_2(k) + 7(2)^{3-k}u(k)$$
$$x_2(k+1) = -\tfrac{3}{2}x_1(k) - 19(2)^{2-k}u(k)$$

Thus

$$\mathbf{A} = \begin{bmatrix} \frac{5}{2} & 1 \\ -\frac{3}{2} & 0 \end{bmatrix}, \qquad \mathbf{B}(k) = \begin{bmatrix} 7(2)^{3-k} \\ -19(2)^{2-k} \end{bmatrix}, \qquad \mathbf{C} = [1 \quad 0]$$

Only the matrix $\mathbf{B}(k)$ is nonstationary. To find the weighting sequence, we write, using (2.39),

$$\mathbf{H}(k, k-j) = \mathbf{C}\mathbf{A}^{k-j-1}\mathbf{B}(j)$$
$$= [1 \quad 0]\begin{bmatrix} \frac{5}{2} & 1 \\ -\frac{3}{2} & 0 \end{bmatrix}^{k-j-1}\begin{bmatrix} 7(2)^{3-j} \\ -19(2)^{2-j} \end{bmatrix}$$

This form is easily simplified by converting to normal coordinates. The eigenvalues of $\mathbf{A}$ are $\lambda_1 = 1$ and $\lambda_2 = \frac{3}{2}$. Hence the eigenvectors are

$$\mathbf{q}^1 = \begin{bmatrix} 1 \\ -\frac{3}{2} \end{bmatrix}, \qquad \mathbf{q}^2 = \begin{bmatrix} 1 \\ -1 \end{bmatrix}$$

and we obtain for the diagonalizing transformation

$$\mathbf{Q} = \begin{bmatrix} 1 & 1 \\ -\frac{3}{2} & -1 \end{bmatrix} \quad \text{and} \quad \mathbf{Q}^{-1} = \begin{bmatrix} -2 & -2 \\ 3 & 2 \end{bmatrix}$$

Thus the normal–coordinate state equation matrices are

$$\mathcal{A} = \mathbf{Q}^{-1}\mathbf{A}\mathbf{Q} = \begin{bmatrix} 1 & 0 \\ 0 & \frac{3}{2} \end{bmatrix}, \qquad \mathcal{B}(k) = \mathbf{Q}^{-1}\mathbf{B}(k) = \begin{bmatrix} 5(2)^{3-k} \\ 2^{4-k} \end{bmatrix}$$
$$\mathcal{C} = \mathbf{C}\mathbf{Q} = [1 \quad 1]$$

Using $\boldsymbol{A}$, $\boldsymbol{B}(k)$, and $\boldsymbol{C}$, we find

$$h(k, k-j) = [5(2)^{k-j-1} + 2(3)^{k-j-1}]2^{4-k}$$

The corresponding transmission matrix is then easily evaluated:

$$\begin{bmatrix} y(1) \\ y(2) \\ y(3) \\ \cdot \\ \cdot \\ \cdot \\ \cdot \\ y(k) \end{bmatrix} = \begin{bmatrix} 56 & 0 & 0 & 0 & \cdots & 0 \\ 64 & 28 & 0 & 0 & & \cdot \\ 76 & 32 & 14 & 0 & & \cdot \\ 94 & 38 & 16 & 7 & & \cdot \\ \cdot & \cdot & \cdot & \cdot & & \cdot \\ \cdot & \cdot & \cdot & \cdot & & \cdot \\ \cdot & \cdot & \cdot & \cdot & & \cdot \\ h(k,k) & h(k,k-1) & \cdot & \cdot & & h(k,1) \end{bmatrix} \begin{bmatrix} u(0) \\ u(1) \\ u(2) \\ \cdot \\ \cdot \\ \cdot \\ \cdot \\ u(k-1) \end{bmatrix}$$

2.7 JORDAN CANONICAL FORM

Every $n \times n$ matrix that is real and symmetric *or* that has n distinct eigenvalues can be diagonalized. Conversely, a matrix that is not real and symmetric *and* that does not have n distinct eigenvalues cannot, in general, be reduced to diagonal form. Fortunately, in systems representing physical dynamic phenomena, the occurrence of multiple eigenvalues is rare; when they do occur, they can frequently be approximated by eigenvalues that differ slightly.[3] In the special circumstances where we wish to preserve the multiplicity of certain eigenvalues, we can reduce the matrix **A** to the so-called *Jordan canonical form.*[4]

Given an $n \times n$ matrix **A** that is not symmetric and that does not have n distinct eigenvalues, the Jordan canonical matrix corresponding to **A** is an $n \times n$ matrix $\mathscr{D}$ for which

$$\begin{aligned} d_{ii} &= \lambda_i \\ d_{i,i+1} &= 1 \quad \text{if } \lambda_i = \lambda_{i+1} \\ &= 0 \quad \text{if } \lambda_i \neq \lambda_{i+1} \\ d_{ij} &= 0 \quad \text{for all } j \neq i \quad \text{and} \quad j \neq i+1, \end{aligned} \tag{2.85}$$

where the λ_i $(i = 1, 2, \ldots, n)$ are the not necessarily distinct eigenvalues of **A**, and where identical eigenvalues are grouped together.

In general, a multiply occurring eigenvalue will yield only one independent eigenvector, in which case the form (2.85) applies. If, however,

[3] Bellman *op. cit.*, p. 198.

[4] Bellman, *op. cit.*, p. 191.

a multiply occurring eigenvalue yields more than one independent eigenvector, (2.85) must be modified so that one element $d_{i,i+1}$ is set equal to zero for each such additional independent eigenvector (even though $\lambda_i = \lambda_{i+1}$).

As an example, if a 5×5 matrix that is not symmetric has only three distinct eigenvalues λ_1, λ_2, and λ_3, where λ_1 occurs with multiplicity three, the corresponding Jordan canonical matrix has the form

$$\begin{bmatrix} \lambda_1 & 1 & 0 & 0 & 0 \\ 0 & \lambda_1 & 1 & 0 & 0 \\ 0 & 0 & \lambda_1 & 0 & 0 \\ 0 & 0 & 0 & \lambda_2 & 0 \\ 0 & 0 & 0 & 0 & \lambda_3 \end{bmatrix}$$

where we have assumed that only one independent eigenvector is associated with λ_1.

Example

Consider the matrix

$$\mathbf{A} = \begin{bmatrix} 0 & 1 & 0 \\ 0 & 0 & 1 \\ 12 & -16 & 7 \end{bmatrix}$$

Upon solving $|\mathbf{A} - \lambda\mathbf{I}| = 0$, we find

$$\lambda = 2, 2, 3$$

Hence, the Jordan canonical matrix is

$$\mathscr{D} = \begin{bmatrix} 2 & 1 & 0 \\ 0 & 2 & 0 \\ 0 & 0 & 3 \end{bmatrix}$$

To find the transformation $\mathbf{Q}$ such that

$$\mathscr{D} = \mathbf{Q}^{-1}\mathbf{A}\mathbf{Q}$$

we solve the equation

$$\mathbf{Q}\mathscr{D} = \mathbf{A}\mathbf{Q}$$

For the arbitrary choice of $q_{11} = 2$, $q_{22} = 3$, and $q_{33} = 4$, the solution obtained is

$$\mathbf{Q} = \begin{bmatrix} 2 & \frac{1}{2} & \frac{4}{9} \\ 4 & 3 & \frac{4}{3} \\ 8 & 10 & 4 \end{bmatrix}$$

2.8 CONDITIONS FOR CONTROLLABILITY AND OBSERVABILITY

When a system is represented in normal form, that is, when the unit-transition matrix is a diagonal matrix, the dynamic modes of the system are isolated from each other. Instead of (2.83) we may write

$$x_i(k+1) = \lambda_i(k)x_i(k) + \sum_{j=1}^{m} b_{ij}(k)u_j(k), \qquad i = 1, 2, \ldots, n \tag{2.86}$$

and

$$y_l(k) = \sum_{i=1}^{n} c_{li}(k)x_i(k), \qquad l = 1, 2, \ldots, p \tag{2.87}$$

It is apparent from (2.86) that if $\boldsymbol{B}(k)$ has an *all-zero row*, that is, if $b_{\rho j}(k) = 0$ for all k, all j and some integer ρ, $1 \leq \rho \leq n$, then the state variable x_ρ cannot be affected, directly or indirectly, by any input. In accordance with the definitions introduced in Section 1.7, we say that under this condition the state variable x_ρ is *uncontrollable.*

In an analogous manner we conclude that a state variable is *unobservable* if there exists some integer μ, $1 \leq \mu \leq n$, such that $c_{l\mu}(k) = 0$ for all k and all l, that is, if $\boldsymbol{C}(k)$ has an *all-zero column.*

The absence of an all-zero row in $\boldsymbol{B}$ and an all-zero column in $\boldsymbol{C}$ are the necessary and sufficient conditions[5] for controllability and observability, respectively, of a linear system all of whose eigenvalues are distinct.

If a system's eigenvalues are not distinct, we must use the Jordan canonical matrix $\mathscr{D}$ instead of the diagonal matrix $\boldsymbol{\Lambda}$. The foregoing conditions for controllability and observability then require some modification, as can be seen from the following example.

Let

$$\mathscr{D} = \begin{bmatrix} \lambda_1 & 1 & 0 & 0 \\ 0 & \lambda_1 & 1 & 0 \\ 0 & 0 & \lambda_1 & 0 \\ 0 & 0 & 0 & \lambda_2 \end{bmatrix}, \qquad \boldsymbol{B} = \begin{bmatrix} b_{11} & b_{12} \\ b_{21} & b_{22} \\ b_{31} & b_{32} \\ b_{41} & b_{42} \end{bmatrix},$$

$$\boldsymbol{C} = \begin{bmatrix} c_{11} & c_{12} & c_{13} & c_{14} \\ c_{21} & c_{22} & c_{23} & c_{24} \end{bmatrix}$$

[5] E. G. Gilbert, "Controllability and Observability in Multivariable Control Systems," *J. Soc. Ind. Appl. Math., Ser. A: Control*, vol. 1, no. 2, pp. 128–151, 1963.

The system, is of course, controllable and observable if $\boldsymbol{B}$ has no all-zero rows and $\boldsymbol{C}$ has no all-zero columns. Although this is a sufficient condition, it is now not a necessary condition. By noting the coupling among the dynamic modes corresponding to the nondistinct eigenvalues, we find that the system is controllable even if the first two rows of $\boldsymbol{B}$ are all-zero. Furthermore, the system is observable even if columns 2 and 3 of $\boldsymbol{C}$ are all-zero. The extension of the conditions for controllability and observability to systems possessing nondistinct eigenvalues is thus a relatively simple matter.

Let us now consider a stationary, linear system that has been transformed into normal form (or Jordan canonical form). Its weighting sequence matrix is given by (2.33) as

$$\mathbf{H}(k - r) = \boldsymbol{C}\boldsymbol{A}^{k-r-1}\boldsymbol{B}, \qquad k > r \tag{2.88}$$

If the system has any dynamic modes that are either not controllable (isolated from input) or not observable (isolated from output), these modes will not contribute to the formation of $\mathbf{H}(k - r)$. *The weighting sequence matrix thus represents only the controllable and observable subsystem of a given system.* A number of important conclusions can be drawn from this.

If a system is described in terms of relations between its input and output vectors, it can always be represented by means of state equations as a completely controllable and observable system. This applies, for example, to all systems that can be described by the difference equation (2.47). Such systems will have precisely as many elementary delay units ("memory cells") as the order of the difference equation or the number of dynamic modes in the weighting sequence matrix. If inadvertently a system realization is obtained which is of an order that is greater than the order of the difference equation, the extraneous state variables can (and should) always be eliminated.

Sometimes, however, a system will not be specified simply in terms of an input–output relation. Sufficient information about the internal structure of the system may be available so that the energy storage elements (corresponding to the "memory cells") can be specifically identified. In this case the order of the system is determined by the total number of storage elements present. The weighting sequence matrix is now possibly a very incomplete description of such a system. The system may possess dynamic modes that are either not controllable or not observable, or both. If this is the case, and if the system properly models a physical dynamic phenomenon, the available inputs and outputs are not adequate to characterize the dynamic behavior of the phenomenon. Such systems require that they be represented by means of state equations (or state transition diagrams).

Example

We wish to determine whether the system shown in Fig. 2-5 is completely controllable and completely observable. We write

$$x_1(k+1) = x_2(k) + 7u(k)$$
$$x_2(k+1) = x_3(k) + 3u(k)$$
$$x_3(k+1) = \tfrac{1}{24}x_1(k) - \tfrac{9}{24}x_2(k) + \tfrac{26}{24}x_3(k) + \tfrac{4}{3}u(k)$$
$$y(k) = 3x_1(k) - 17x_2(k) + 24x_3(k)$$

Hence

$$\mathbf{A} = \begin{bmatrix} 0 & 1 & 0 \\ 0 & 0 & 1 \\ \frac{1}{24} & -\frac{9}{24} & \frac{26}{24} \end{bmatrix}, \qquad \mathbf{B} = \begin{bmatrix} 7 \\ 3 \\ \frac{4}{3} \end{bmatrix},$$

$$\mathbf{C} = [3 \quad -17 \quad 24]$$

Solving $|\mathbf{A} - \lambda\mathbf{I}| = 0$, we find that the eigenvalues are $\frac{1}{2}$, $\frac{1}{3}$, and $\frac{1}{4}$. For the diagonalizing transformation we use

$$\mathbf{Q} = \begin{bmatrix} 4 & 9 & 16 \\ 2 & 3 & 4 \\ 1 & 1 & 1 \end{bmatrix} \quad \text{and} \quad \mathbf{Q}^{-1} = \begin{bmatrix} \frac{1}{2} & -\frac{7}{2} & 6 \\ -1 & 6 & -8 \\ \frac{1}{2} & -\frac{5}{2} & 3 \end{bmatrix}$$

Then

$$\mathcal{A} = \mathbf{Q}^{-1}\mathbf{A}\mathbf{Q} = \begin{bmatrix} \frac{1}{2} & 0 & 0 \\ 0 & \frac{1}{3} & 0 \\ 0 & 0 & \frac{1}{4} \end{bmatrix}, \qquad \mathcal{B} = \mathbf{Q}^{-1}\mathbf{B} = \begin{bmatrix} 1 \\ \frac{1}{3} \\ 0 \end{bmatrix},$$

$$\mathcal{C} = \mathbf{C}\mathbf{Q} = [2 \quad 0 \quad 4]$$

The system's eigenvalues are all distinct. Since $\mathcal{b}_3 = 0$, the dynamic mode represented by the eigenvalue $\frac{1}{4}$ is not controllable. Similarly, since $\mathcal{c}_2 = 0$, the dynamic mode corresponding to the eigenvalue $\frac{1}{3}$ is not

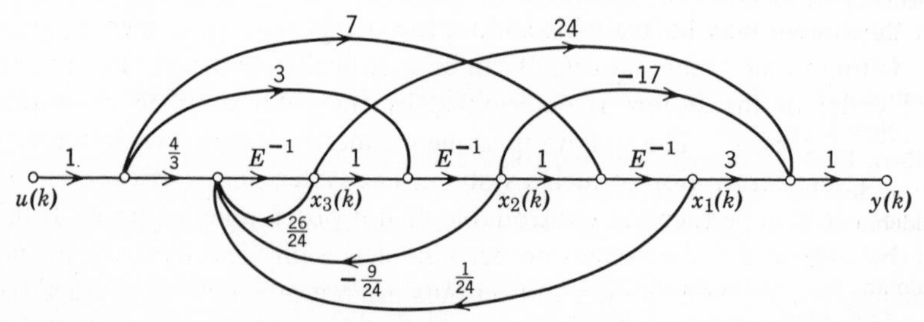

Figure 2-5 State transition diagram for system of example. (*Note:* E^{-1} denotes the unit-delay operator.)

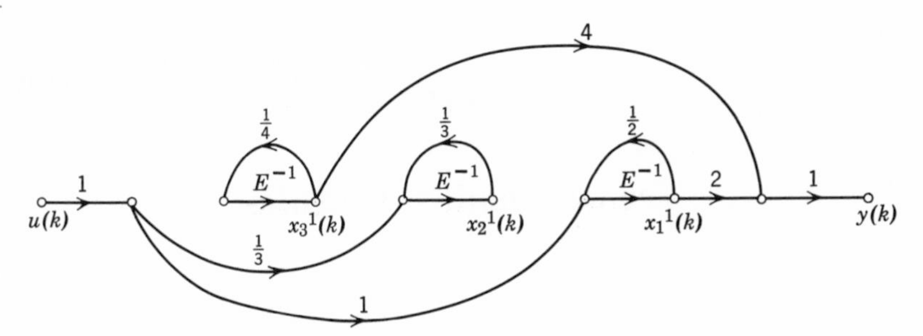

Figure 2-6 Normal-coordinate state-transition diagram for system of example.

observable. The system is thus neither completely controllable nor completely observable.

We can redraw the state-transition diagram of Fig. 2-5 using the transformed state coordinates $\mathbf{x}^1 = \mathbf{Q}^{-1}\mathbf{x}$. This is shown in Fig. 2-6. The controllability and observability properties of the system now become clearly evident.

The weighting sequence matrix is obtained by using (2.88):

$$\mathbf{H}(k - r) = \boldsymbol{C}\boldsymbol{A}^{k-r-1}\boldsymbol{B} = 4(\tfrac{1}{2})^{k-r} \quad \text{for } k > r$$

As expected, the weighting sequence matrix describes only the one dynamic mode of the system that is both controllable and observable.

REFERENCES

Ayres, Frank, *Theory and Problems of Matrices*, Schaum Publishing Co., New York, 1962.

Bellman, R., *Introduction to Matrix Analysis*, McGraw-Hill Book Co., New York, 1960.

Bertram, J. E., "The Concept of State in the Analysis of Discrete-Time Control Systems," *Proc. Joint Autom. Control Conf.*, American Institute of Electrical Engineers, pp. 11-1-1 to 11-1-7, June 1962.

Coddington, E. A. and N. Levinson, *Theory of Ordinary Differential Equations*, McGraw-Hill Book Co., New York, 1955.

Friedland, B., *Theory of Time-Varying Sampled-Data Systems*, Tech. Report T-19/B, Department of Electrical Engineering, Columbia University, New York, April 1957.

Gantmacher, F. R., *Applications of the Theory of Matrices*, Interscience Publishers, New York, 1959.

Gilbert, E. G., "Controllability and Observability in Multivariable Control Systems," *J. Soc. Ind. Appl. Math., Ser. A: Control*, vol. 1, no. 2, pp. 128–151, 1963.

Hildebrand, F. B., *Methods of Applied Mathematics*, Prentice-Hall, Englewood Cliffs, N.J., 1952.

Kalman, R., "Mathematical Description of Linear Dynamical Systems," *J. Soc. Ind. Appl. Math., Ser. A: Control*, vol. 1, no. 2, pp. 152–192, 1963.

Levy, H. and F. Lessman, *Finite Difference Equations*, The Macmillan Co., New York, 1961.

CHAPTER 3

Transformation Calculus

3.1 INTRODUCTION

In this chapter we shall concern ourselves with a transformation calculus for discrete-time functions. This calculus is based on a transformation that permits the replacement of the convolution summation operation in the time domain by a simple multiplication in the transformed domain. It serves the discrete-time domain in the same manner that the familiar Laplace transform serves the continuous-time domain.

The transformation calculus is primarily of value for the solution of problems involving linear, stationary systems. It provides an important body of techniques that supplement the time-domain techniques described in Chapter 2. Our approach will be quite general. Although for purposes of system analysis we are nearly always restricted to time sequences that vanish for negative time, we shall not assume this restriction here. The calculus can be advantageously applied to the solution of a variety of problems for which the independent parameter is not time and hence is not restricted to positive values. It is, therefore, desirable to develop the transformation calculus to include sequences that are nonzero for both positive and negative values of the independent parameter.

3.2 THE GENERATING FUNCTION

Consider the time sequence $\{f(k)\}$ which assumes the values $f(a)$, $f(a+1), \ldots, f(b)$ (a and b integers, $b > a$) as k ranges over the integers in the interval a to b. We define a new function $F(z)$ as a power series in z^{-n} having as coefficients the values of the time sequence $\{f(k)\}$. This new function will be called the *generating function* of $\{f(k)\}$ and will be written as[1]

$$F(z)_{a,b} = f(a)z^{-a} + f(a+1)z^{-a-1} + \cdots + f(b)z^{-b} \tag{3.1}$$

[1] Most texts in the calculus of finite differences prefer a series in positive powers of z for positive k. Although this is unquestionably preferable, it is in contradiction to the now widespread practice in engineering to use a transformation based on negative powers of z for positive k.

or

$$\boxed{\mathscr{G}[f(k)]_{a,b} = F(z)_{a,b} = \sum_{k=a}^{b} f(k)z^{-k}} \tag{3.2}$$

The function $\{f(k)\}$ is also referred to as the *determining function* of $F(z)$.[2]

There is no restriction on the limits a and b, which describe the interval over which $\{f(k)\}$ is defined. In the most general case, the interval is $-\infty \leq k \leq \infty$. We note, however, that if either or both of the limits in (3.2) are permitted to become infinite, restrictions must be placed both on the nature of $\{f(k)\}$ and on the values that z may assume if the summation is to converge. A number of different cases will now be examined.

1. If $a \leq k \leq b$, a and b finite, there is no restriction on $\{f(k)\}$, and z may assume all values except $z = \infty$ if $a < 0$, and except $z = 0$ if $b > 0$.

2. If $a \leq k \leq \infty$, $\{f(k)\}$ must be of exponential order; that is, its rate of increase as $k \to \infty$ may be no greater than that of a geometric series. Hence as $k \to \infty$, $|f(k)|^{1/k}$ will approach one or more finite, positive limit points.[3] Let us denote the *greatest* of these limit points by $\overline{\lim\limits_{k\to\infty}}\,|f(k)|^{1/k}$. If we now let

$$\overline{\lim_{k\to\infty}}\,|f(k)|^{1/k} = R_c \tag{3.3}$$

then the series $F(z) = \sum_{k=a}^{\infty} f(k)z^{-k}$ will converge absolutely for all z such that

$$|z| > R_c \tag{3.4}$$

with the exception of $z = \infty$ if $a < 0$. The number R_c is known as the *radius of convergence* of the series $\sum_{k=a}^{\infty} f(k)z^{-k}$. If z is a complex variable, $z = u + jv$, the series will converge absolutely for all points in the z-plane that lie *outside* the circle which has a radius R_c and is centered at the origin (with the possible exception of the point $z = \infty$). The series will diverge for all points inside this circle. (No definite statement about convergence can be made for points on the circle.)

Example

$$f(k) = 2^k \quad \text{for } k \geq 0, \qquad f(k) = 0 \quad \text{for } k < 0$$

$$\therefore R_c = 2$$

$$F(z) = \sum_{k=0}^{\infty} 2^k z^{-k} \quad \text{for all } |z| > 2$$

[2] Strictly speaking, the generating function should be denoted by either F or $\{F(z)\}$, and its value, at z, by $F(z)$. To comply with common usage, however, we shall normally use $F(z)$ to denote both the function as well as its value at z.

[3] K. Knopp, *Infinite Sequences and Series*, Dover Publications, New York, 1956.

3. If $-\infty \leq k \leq b$, $\{f(k)\}$ must be of exponential order as $k \to -\infty$. Let us denote by $\underline{\lim}_{k\to-\infty} |f(k)|^{1/k}$ the *smallest* of the finite positive limit points that $|f(k)|^{1/k}$ approaches as $k \to -\infty$.[4]

If we let

$$\underline{\lim}_{k\to-\infty} |f(k)|^{1/k} = R_c \tag{3.5}$$

then the series $F(z) = \sum_{k=-\infty}^{b} f(k)z^{-k}$ will converge absolutely for all z for which

$$|z| < R_c \tag{3.6}$$

with the exception of $z = 0$ if $b > 0$.

Example

$$\begin{aligned} f(k) &= 2^k \quad \text{for even } k \leq 0 \\ &= 3^k \quad \text{for odd } k < 0 \\ &= 0 \quad \text{for } k > 0 \end{aligned}$$

As $k \to -\infty$, the value of $|f(k)|^{1/k}$ will alternate between 2 and 3; that is, there are two distinct limit points. In accordance with (3.5), the smaller limit point must be used. Hence

$$R_c = 2$$

and

$$F(z) = 1 + \tfrac{1}{3}z + \tfrac{1}{4}z^2 + \tfrac{1}{27}z^3 + \tfrac{1}{16}z^4 + \cdots \quad \text{for all } |z| < 2$$

4. If $-\infty \leq k \leq \infty$, $\{f(k)\}$ must be of exponential order such that two finite positive numbers, R_{c+} and R_{c-}, can be defined as follows:

$$\begin{aligned} \overline{\lim}_{k\to\infty} |f(k)|^{1/k} &= R_{c+} \\ \underline{\lim}_{k\to-\infty} |f(k)|^{1/k} &= R_{c-} \end{aligned} \tag{3.7}$$

The series $F(z) = \sum_{k=-\infty}^{\infty} f(k)z^{-k}$ will converge absolutely for all z for which

$$R_{c+} < |z| < R_{c-} \tag{3.8}$$

If $z = u + jv$, the region of convergence of $F(z)$ consists of all the points in the z plane which lie within the annulus bounded by the circles, both centered at the origin, of radius R_{c+} and R_{c-}. We observe that for $F(z)$ to converge, it is necessary that $R_{c-} > R_{c+}$. This condition is *always* satisfied if $f(k)$ tends to 0 as $k \to +\infty$ *and* as $k \to -\infty$. It is of interest to note, however, that the sequence $\{f(k)\}$ may diverge as $k \to \infty$ or $k \to -\infty$ (but not both!), provided the rate of exponential increase in the direction

[4] If there is only one limit point as $k \to -\infty$ (or as $k \to \infty$), the question as to which is the smallest (or greatest) will, of course, not arise.

of k for which $\{f(k)\}$ diverges is less than the rate of exponential decrease in the direction of k for which $\{f(k)\}$ tends to zero. (It *must* tend to zero in at least one of the two directions, $k \to \infty$ or $k \to -\infty$, for a region of convergence to exist.)

Example 1

$$f(k) = 2^k \quad \text{for } k \geq 0,$$
$$= 3^k \quad \text{for } k \leq 0$$
$$\therefore \quad R_{c+} = 2$$
$$R_{c-} = 3$$
$$F(z) = \sum_{k=0}^{\infty} 2^k z^{-k} + \sum_{k=-\infty}^{-1} 3^k z^{-k} \quad \text{for } 2 < |z| < 3$$

The region of convergence for this example is shown in Fig. 3-1.

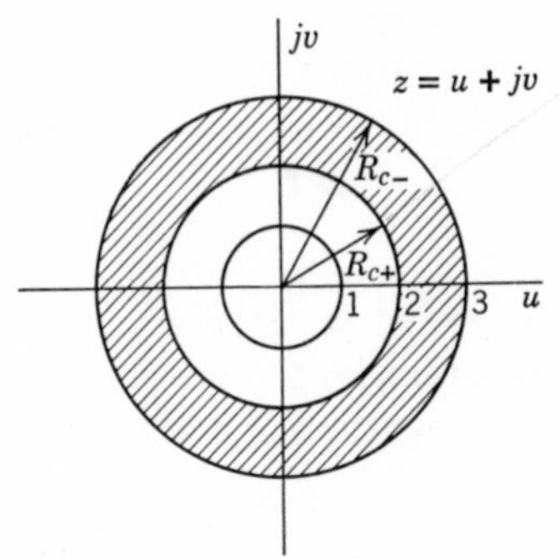

Figure 3-1 Region of convergence of $F(z)$ in plane $z = u + jv$. $F(z)$ converges for all z in shaded area.

Example 2

$$f(k) = 3^k \quad \text{for } k \geq 0, \qquad f(k) = 2^k \quad \text{for } k \leq 0$$
$$\therefore \quad R_{c+} = 3$$
$$R_{c-} = 2$$

$R_{c+} > R_{c-}$ and hence there is no region in the z plane for which $F(z)$ converges.

It is apparent from (3.2) that the relation between a generating function $F(z)$ and its determining function $\{f(k)\}$ is that of a linear operation, subject, however, to the satisfaction of the applicable convergence requirement. Thus if

$$F_1(z) = \mathscr{G}[f_1(k)] \quad \text{for all } |z| > R_{c1}$$
$$F_2(z) = \mathscr{G}[f_2(k)] \quad \text{for all } |z| > R_{c2}$$

then

$$\mathscr{G}[a_1 f_1(k) + a_2 f_2(k)] = a_1 F_1(z) + a_2 F_2(z) \tag{3.9}$$

for all z *in the region of convergence common to both* $F_1(z)$ *and* $F_2(z)$, *and for all arbitrary constants* a_1 *and* a_2.

3.3 TRANSFORMATION OF THE CONVOLUTION SUMMATION

Consider a linear, stationary system that is subjected to an input sequence $\{u(k)\}$, $-\infty \leq k \leq \infty$. If the system is not restricted to be nonanticipatory, its weighting sequence $\{h(k)\}$ may be nonzero for all k, $-\infty \leq k \leq \infty$. Generalizing (2.6), we have

$$y(k) = \sum_{i=-\infty}^{\infty} h(k-i)u(i) \tag{3.10}$$

We assume that a region of convergence exists for $Y(z)$ and write, using (3.2) with $a = -\infty$ and $b = \infty$,

$$Y(z) = \mathscr{G}[y(k)] = \sum_{k=-\infty}^{\infty} \sum_{i=-\infty}^{\infty} h(k-i)u(i)z^{-k} \tag{3.11}$$

A power series that converges absolutely also converges uniformly. Accordingly, the order of summation in (3.11) may be reversed without changing the value of the sum:[5]

$$Y(z) = \sum_{i=-\infty}^{\infty} u(i)z^{-i} \sum_{k=-\infty}^{\infty} h(k-i)z^{-(k-i)} \tag{3.12}$$

Let $j = k - i$. Then

$$Y(z) = \sum_{i=-\infty}^{\infty} u(i)z^{-i} \sum_{j=-\infty}^{\infty} h(j)z^{-j} \tag{3.13}$$

$$= U(z)H(z) \tag{3.14}$$

where the $U(z)$ and $H(z)$ are simply the generating functions of $\{u(k)\}$ and $\{h(k)\}$, respectively, and where the region of convergence of $Y(z)$ is that region of the z-plane in which *both* $U(z)$ and $H(z)$ converge. Expressed another way,

$$\mathscr{G}[y(k)] = \mathscr{G}[u(k)]\mathscr{G}[h(k)] \tag{3.15}$$

provided a common region of convergence exists.

We note that the relation expressed by the convolution summation (2.6) becomes a simple multiplication when the variables are transformed into generating functions. It is this property that makes the generating function a powerful instrument for dealing with functions of a discrete variable. Thus to find the output y of a system h subjected to an input u, we form the generating functions $H(z)$ and $U(z)$, multiply them, and then perform an inverse transformation on the product to obtain

$$y(k) = \mathscr{G}^{-1}[U(z)H(z)] \tag{3.16}$$

Alternately, given a linear stationary system, if an input sequence $\{u(k)\}$ results in an output sequence $\{y(k)\}$, we may define a *system transfer function* $H(z)$ where

$$H(z) = \frac{Y(z)}{U(z)} \tag{3.17}$$

[5] E. C. Titchmarsh, *The Theory of Functions* (2nd ed.), Oxford University Press, London, 1939.

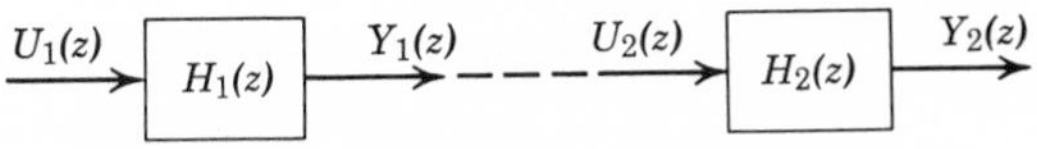

$$U_1(z) \rightarrow \boxed{H_1(z)} \rightarrow Y_1(z) - - - U_2(z) \rightarrow \boxed{H_2(z)} \rightarrow Y_2(z)$$

Figure 3-2 Two systems connected in cascade.

If we are given two systems arranged so that the output variable of one becomes the input variable of the other (see Fig. 3-2), where

$$H_1(z) = \frac{Y_1(z)}{U_1(z)} \tag{3.18}$$

$$H_2(z) = \frac{Y_2(z)}{U_2(z)} \tag{3.19}$$

and

$$Y_1(z) = U_2(z)$$

then

$$Y_2(z) = H_2(z)H_1(z)U_1(z) \tag{3.20}$$

or

$$H_0(z) = \frac{Y_2(z)}{U_1(z)} = H_2(z)H_1(z) \tag{3.21}$$

that is, two linear, stationary systems connected in cascade are characterized by a transfer function that is equal to the product of their individual transfer functions.

3.4 THE $\mathcal{Z}$ TRANSFORM

The generating function of f, where $f(k) = 0$ for $k < 0$, is also known as the *z transform* of f and written as $\mathcal{Z}[f(k)]$. Although the name *generating function* antedates that of z transform, the latter has gained wide acceptance in the field of system analysis. In compliance with this usage, the term generating function will be reserved for discussions pertaining to the general mathematical aspects of the transformation, while the term z transform will be used in connection with system analysis problems for $k \geq 0$. The former may be applied anywhere over the interval $-\infty \leq k \leq \infty$; the latter will be restricted to functions which vanish for $k < 0$.[6] Thus

$$\mathcal{G}[f(k)] = \sum_{k=-\infty}^{\infty} f(k)z^{-k} \tag{3.22}$$

and

$$\mathcal{Z}[f(k)] = \sum_{k=0}^{\infty} f(k)z^{-k} \tag{3.23}$$

[6] Some authors use the term z transform exclusively and define a "bilateral z transform" to extend the transformation to functions which do not vanish for $k < 0$.

3.5 DETERMINATION OF $\mathcal{Z}$ TRANSFORMS

A variety of methods exists for finding closed expressions for the z transforms of common time sequences. The simplest method utilizes direct summation. Thus if $f(k) = a^k$ for $k \geq 0$,

$$\mathcal{Z}[f(k)] = F(z) = \sum_{k=0}^{\infty} a^k z^{-k} \tag{3.24}$$

Using the well-known formula for the sum of a geometric series,

$$S = A\frac{1 - r^k}{1 - r}$$

where $r = az^{-1}$, $A = 1$, and $k \rightarrow \infty$, we obtain

$$\boxed{\mathcal{Z}[a^k] = \frac{1}{1 - az^{-1}} \quad \text{for } |z| > |a|} \tag{3.25}$$

Letting $a = 1$, we obtain the z transform of the constant sequence $f(k) = 1$ for $k \geq 0$:

$$\boxed{\mathcal{Z}[1] = \frac{1}{1 - z^{-1}} \quad \text{for } |z| > 1} \tag{3.26}$$

Similarly, by letting $a = e^{j\omega}$,

$$\begin{aligned} \mathcal{Z}[e^{j\omega k}] &= \frac{1}{1 - e^{j\omega}z^{-1}} \quad \text{for } |z| > 1 \\ &= \frac{1}{1 - z^{-1}\cos\omega - jz^{-1}\sin\omega} \\ &= \frac{1 - z^{-1}\cos\omega + jz^{-1}\sin\omega}{1 - 2z^{-1}\cos\omega + z^{-2}} \end{aligned} \tag{3.27}$$

If we now equate the real and imaginary parts separately,

$$\boxed{\mathcal{Z}[\sin\omega k] = \frac{z^{-1}\sin\omega}{1 - 2z^{-1}\cos\omega + z^{-2}} \quad \text{for } |z| > 1} \tag{3.28}$$

$$\boxed{\mathcal{Z}[\cos\omega k] = \frac{1 - z^{-1}\cos\omega}{1 - 2z^{-1}\cos\omega + z^{-2}} \quad \text{for } |z| > 1} \tag{3.29}$$

We may find additional z transforms by differentiating or integrating known z transforms within their region of convergence. Since the z transforms are power series, they converge uniformly within their region of absolute convergence and may be differentiated or integrated an arbitrary number of times. The resulting series will possess the same region of convergence.[7]

The method of differentiating a known z transform to obtain an additional z transform relation is illustrated in the following derivation:

Given

$$F(z) = \sum_{k=0}^{\infty} f(k)z^{-k} \quad \text{for } |z| > R_c \tag{3.30}$$

then

$$\frac{d}{dz} F(z) = -\sum_{k=0}^{\infty} kf(k)z^{-k-1}$$

$$-z\frac{d}{dz} F(z) = \sum_{k=0}^{\infty} kf(k)z^{-k}$$

and

$$\mathcal{Z}[kf(k)] = -z\frac{d}{dz} F(z) \quad \text{for } |z| > R_c \tag{3.31}$$

More generally, for any integer $\nu \geq 0$,

$$\mathcal{Z}[k^{\nu}f(k)] = \left(-z\frac{d}{dz}\right)^{\nu} F(z) \quad \text{for } |z| > R_c \tag{3.32}$$

where the expression $\left(-z\frac{d}{dz}\right)^{\nu}$ is to be interpreted as the ν-fold iteration of the operation $-z\frac{d}{dz}$ and where the region of convergence is the same as that of $F(z)$.

If we let $f(k) = 1$ in (3.31) and substitute the result of (3.26),

$$\mathcal{Z}[k] = -z\frac{d}{dz}\left(\frac{1}{1 - z^{-1}}\right)$$

$$\mathcal{Z}[k] = \frac{z^{-1}}{(1 - z^{-1})^2} \quad \text{for } |z| > 1 \tag{3.33}$$

[7] Titchmarsh, *op. cit.*, Section 1.73.

Using (3.32) with $\nu = 2$, we have

$$\mathcal{Z}[k^2] = \frac{z^{-1}(1 + z^{-1})}{(1 - z^{-1})^3} \quad \text{for } |z| > 1 \tag{3.34}$$

The integration of z transforms may also be employed to find additional z transforms. Thus multiplying both sides of (3.26) by z^{-2} gives

$$\sum_{k=0}^{\infty} z^{-k-2} = \frac{z^{-2}}{1 - z^{-1}} \quad \text{for } |z| > 1$$

Integrating with respect to z, we have

$$\sum_{k=0}^{\infty} \frac{z^{-k-1}}{-k-1} = \log(1 - z^{-1}) \tag{3.35}$$

Hence

$$\sum_{k=1}^{\infty} \frac{z^{-k}}{k} = -\log(1 - z^{-1}) \quad \text{for } |z| > 1 \tag{3.36}$$

and

$$\mathcal{Z}\left[\frac{1}{k}\right] = -\log(1 - z^{-1}), \qquad k \geq 1, |z| > 1 \tag{3.37}$$

Finally, we may make use of any known formula for the sum of a convergent power series. Thus using the relation

$$e^x = \sum_{k=0}^{\infty} \frac{x^k}{k!} \quad \text{for } |x| < \infty \tag{3.38}$$

we obtain (by replacing x by z^{-1})

$$\mathcal{Z}\left[\frac{1}{k!}\right] = e^{1/z} \quad \text{for } |z| > 0 \tag{3.39}$$

Given that $\mathcal{Z}[f(k)] = F(z)$ for $|z| > R_c$, $f(k) = 0$ for $k < 0$, then

$$\begin{aligned} \mathcal{Z}[f(k+1)] &= \sum_{k=0}^{\infty} f(k+1) z^{-k} \\ &= z \sum_{k=0}^{\infty} f(k+1) z^{-k-1} = z\left[\sum_{k=0}^{\infty} f(k) z^{-k} - f(0)\right] \end{aligned} \tag{3.40}$$

$$\mathcal{Z}[f(k+1)] = z[F(z) - f(0)] \quad \text{for } |z| > R_c \tag{3.41}$$

In general, for $m \geq 0$,

$$\mathcal{Z}[f(k+m)] = z^m\left[F(z) - \sum_{i=0}^{m-1} f(i)z^{-i}\right] \quad \text{for } |z| > R_c \tag{3.42}$$

Since $f(k) = 0$ for $k < 0$,

$$\mathcal{Z}[f(k-m)] = z^{-m}F(z) \quad \text{for } |z| > R_c \tag{3.43}$$

For the *forward-difference operator* Δ, defined by

$$\Delta f(k) = f(k+1) - f(k)$$

we have from (3.41)

$$\mathcal{Z}[\Delta f(k)] = (z-1)F(z) - zf(0) \quad \text{for } |z| > R_c \tag{3.44}$$

Similarly, for the *backward-difference operator* ∇, defined by

$$\nabla f(k) = f(k) - f(k-1)$$

we have from (3.43)

$$\mathcal{Z}[\nabla f(k)] = (1 - z^{-1})F(z) \quad \text{for } |z| > R_c \tag{3.45}$$

The following relation is of considerable importance:

$$\mathcal{Z}[a^k f(k)] = \sum_{k=0}^{\infty} a^k f(k) z^{-k}$$

$$= \sum_{k=0}^{\infty} f(k)(a^{-1}z)^{-k}$$

for

$$|z| > \overline{\lim_{k\to\infty}}\, |a|\, |f(k)|^{1/k}$$

Thus if $\mathcal{Z}[f(k)] = F(z)$ for $|z| > R_c$, then

$$\mathcal{Z}[a^k f(k)] = F(a^{-1}z) \quad \text{for } |z| > |a|\, R_c \tag{3.46}$$

that is, if we are given the z transform of an arbitrary sequence $\{f(k)\}$, we can immediately determine the z transform of the sequence $\{a^k f(k)\}$.

Example

Consider a system described by the difference equation:

$$y(k+2) - 2.5y(k+1) + 1.5y(k) = u(k)$$

The initial conditions are $y(0)$ and $y(1)$. The input $u(k) = 2^k$ is applied at $k = 0$. We desire $y(k)$ for $k \geq 2$.

Using (3.42), we obtain the z transform of the equation:

$$z^2[Y(z) - y(0) - y(1)z^{-1}] - 2.5z[Y(z) - y(0)] + 1.5Y(z) = U(z)$$

Combining terms and using (3.25) for $U(z)$, we have

$$(z^2 - 2.5z + 1.5)Y(z) - z(z - 2.5)y(0) - zy(1) = \frac{1}{1 - 2z^{-1}}$$

$$Y(z) = \frac{z}{(z - 2)(z - 1)(z - 1.5)} + \frac{z(z - 2.5)y(0) + zy(1)}{(z - 1)(z - 1.5)}, \qquad |z| > 2$$

The time sequence corresponding to $Y(z)$ is the desired output sequence. It is obtained by taking the inverse transformation of $Y(z)$. (Methods for performing the inverse transformation are discussed in the next section.) Note that the eigenvalues of the system are the roots of the equation

$$z^2 - 2.5z + 1.5 = 0$$

that is, $z = 1$ and $z = 1.5$. The transformation calculus thus provides a very convenient means for rapidly obtaining the eigenvalues of a system.

In the preceding discussions the transformation calculus was applied only to single time sequences. The calculus can, however, be applied equally well to vector time sequences. The result is then a vector generating function (or vector z transform). As an illustration, consider the state equation (2.26) for a stationary system:

$$\mathbf{x}(k + 1) = \mathbf{A}\mathbf{x}(k) + \mathbf{B}\mathbf{u}(k) \tag{2.26}$$

Let $\mathbf{X}(z)$ and $\mathbf{U}(z)$ be the z transforms of $\mathbf{x}(k)$ and $\mathbf{u}(k)$, respectively. Then from (3.41),

$$z\mathbf{X}(z) - z\mathbf{x}(0) = \mathbf{A}\mathbf{X}(z) + \mathbf{B}\mathbf{U}(z)$$
$$[z\mathbf{I} - \mathbf{A}]\mathbf{X}(z) = z\mathbf{x}(0) + \mathbf{B}\mathbf{U}(z)$$

Premultiplying both sides of this equation by $[z\mathbf{I} - \mathbf{A}]^{-1}$ then yields

$$\begin{aligned}\mathbf{X}(z) &= [z\mathbf{I} - \mathbf{A}]^{-1}z\mathbf{x}(0) + [z\mathbf{I} - \mathbf{A}]^{-1}\mathbf{B}\mathbf{U}(z)\\ &= [\mathbf{I} - z^{-1}\mathbf{A}]^{-1}\mathbf{x}(0) + z^{-1}[\mathbf{I} - z^{-1}\mathbf{A}]^{-1}\mathbf{B}\mathbf{U}(z)\end{aligned} \tag{3.47}$$

The last equation is the vector z transform of $\mathbf{x}(k)$ as given by (2.29). Note that since from (2.28)

$$\mathbf{\Phi}(k) = \mathbf{A}^k \quad \text{for } k \geq 0$$

then

$$\mathcal{Z}[\mathbf{A}^k] = [\mathbf{I} - z^{-1}\mathbf{A}]^{-1} \tag{3.48}$$

This is the vector form of (3.25).

3.6 THE INVERSE TRANSFORMATION

A number of methods exist for finding the sequence $\{f(k)\}$, given the *generating function* $\mathcal{G}[f(k)] = F(z)$ in a specified region of convergence of the z plane.

In the most direct method, $F(z)$ is expanded into a convergent power series in z^{-1}:

$$F(z) = \sum_{k=-\infty}^{\infty} f(k)z^{-k}$$

If $F(z)$ is given in the form of a rational function, this is accomplished simply by dividing the denominator into the numerator. The coefficients of the z^{-k} in the resulting expansion are the desired values $f(k)$ of the time sequence. An expression for the general term can then often be found by inspection.

If $F(z)$ converges for $|z| > R_c$, the division must be performed so as to obtain a convergent infinite series in increasing powers of z^{-1}. (The corresponding series in increasing powers of z does not converge for $|z| > R_c$.)

If $F(z)$ converges for $|z| < R_c$, the division must be performed to yield a convergent infinite series in increasing powers of z.

If $F(z)$ converges in some annulus, $R_{c-} > |z| > R_{c+}$, we must first decompose $F(z)$ into a sum of two functions, $F(z) = F_1(z) + F_2(z)$ such that $F_1(z)$ converges for $|z| > R_{c+}$, and $F_2(z)$ converges for $|z| < R_{c-}$; that is, $F_1(z)$ must contain only those poles of $F(z)$ which are located inside or on the circle of radius R_{c+} in the z plane, and $F_2(z)$ must contain only those poles which are located outside or on the circle of radius R_{c-}. $F_1(z)$ and $F_2(z)$ can then be separately expanded into convergent infinite series.

The foregoing method will yield the values of the desired time sequence in a sequential manner, beginning with the value at (or near) $k = 0$. If $f(k)$ is desired for either a large positive or large negative k, a lengthy calculation is involved.[8] In addition the numerical calculation may be subject to serious error build-up unless special precautions are observed.[9] Except when $F(z)$ is a relatively simple function, it is usually not possible to deduce an expression for the general term from a set of values of the time sequence.

[8] A computational procedure which facilitates the expansion process is given in Appendix A.

[9] W. H. Huggins, "A Low-Pass Transformation for $\mathcal{Z}$ Transforms," *IRE Trans. Circuit Theory*, vol. 1, pp. 69–70, September 1954.

Example

Find $f(k)$, given that

$$F(z) = \frac{7z^2 - 19z}{z^2 - 5z + 6} \quad \text{for all } |z| > 3$$

Since $F(z)$ converges for all $|z| > R_c$, it can be expanded in an infinite series in z^{-k} from $k = 0$ to $k = \infty$. The appropriate division of denominator into numerator yields

$$F(z) = 7 + 16z^{-1} + 38z^{-2} + 94z^{-3} + \cdots \to \infty$$

Hence

$$f(0) = 7, \quad f(1) = 16, \quad f(2) = 38, \ldots \text{ etc.}$$

(With some ingenuity it may be recognized that the general term is given by $f(k) = 5(2)^k + 2(3)^k$ for all $k \geq 0$.)

3.7 INVERSION BY PARTIAL-FRACTION EXPANSION

An alternate method for finding the $f(k)$ that correspond to a given rational $F(z)$ is to make a partial-fraction expansion of $F(z)$ and then to identify each of the terms of the expansion (if necessary with the aid of a table of transform pairs). Since an expansion into partial fractions is a purely algebraic operation, it can be carried out without any consideration to the region of convergence of $F(z)$. Many dfferent forms of partial-fraction expansions are possible, and strictly speaking, it is immaterial which particular form is chosen. However, from a practical point of view, some forms yield more easily recognizable terms than others. If $F(z)$ converges for $|z| > R_c$ and contains only simple poles, inversion by means of a partial-fraction expansion is straightforward. However, if $F(z)$ has high-order multiple poles and converges only in an annular region, the identification of the partial-fraction expansion terms may become extremely difficult. In this section we shall describe a particular form of partial-fraction expansion that leads to easily identifiable terms even in the case of multiple poles and annular convergence.

Consider an $F(z)$ given as the ratio of two polynomials in z:

$$F(z) = \frac{N(z)}{D(z)} = \frac{b_p z^p + b_{p-1} z^{p-1} + \cdots + b_0}{d_q z^q + d_{q-1} z^{q-1} + \cdots + d_0} \tag{3.49}$$

If $p \geq q$, it is necessary to divide $D(z)$ into $N(z)$ until a remainder polynomial $N_1(z)$ is obtained which is of a degree one less than the degree of $D(z)$:

$$F(z) = \sum_{l=0}^{p-q} c_l z^l + \frac{N_1(z)}{D(z)} \tag{3.50}$$

Next we determine the poles of $F(z)$ by factoring $D(z)$. In setting up the form of the partial-fraction expansion, we use only terms having the form given in the left column of Table 3-1.[10] For a pole of multiplicity v, we use the term corresponding to it as well as to all lower-order terms. Thus for every simple pole a_1, the expansion will have a term

$$\frac{c}{z - a_1}$$

and for every triple pole a_2 it will contain the set of terms

$$\frac{c_1 z(z + a_2)}{(z - a_2)^3} + \frac{c_2 z}{(z - a_2)^2} + \frac{c_3}{z - a_2}$$

where the c's are the constants that are to be determined.

TABLE 3-1 ELEMENTARY TRANSFORM TERMS FOR PARTIAL-FRACTION EXPANSION

Elementary Transform Term $F_i(z)$	Corresponding Time Sequence	
	(I) $F_i(z)$ converges for $\|z\| > R_c$	(II) $F_i(z)$ converges for $\|z\| < R_c$
1. $\dfrac{1}{z - a}$	$a^{k-1}\|_{k\geq 1}$	$-a^{k-1}\|_{k\leq 0}$
2. $\dfrac{z}{(z - a)^2}$	$ka^{k-1}\|_{k\geq 1}$	$-ka^{k-1}\|_{k\leq 0}$
3. $\dfrac{z(z + a)}{(z - a)^3}$	$k^2 a^{k-1}\|_{k\geq 1}$	$-k^2 a^{k-1}\|_{k\leq 0}$
4. $\dfrac{z(z^2 + 4az + a^2)}{(z - a)^4}$	$k^3 a^{k-1}\|_{k\geq 1}$	$-k^3 a^{k-1}\|_{k\leq 0}$

Once the forms of the q terms in the partial-fraction expansion have been established, we solve for the unknown coefficients in the customary manner. The identification of the terms is then accomplished by again referring to Table 3-1. If for a pole a_i, $F(z)$ converges for *some* $|z| > a_i$, then the corresponding time sequence is found in column I; if $F(z)$ converges for *some* $|z| < a_i$, then the corresponding time sequence must be obtained from column II.[11]

[10] The table can be extended to poles of multiplicity greater than four by the application of (3.32).

[11] Convergence for *some* $|z|$ in the indicated region is both necessary and sufficient to establish whether an entry is to be selected from column I or column II. $F(z)$ may, of course, converge for *all* $|z|$ in the indicated region; however, this is not necessary.

Example

$$F(z) = \frac{-2z^4 + 18z^3 - 48z^2 + 63z - 54}{z^4 - 8z^3 + 21z^2 - 18z}$$

Find $f(k)$, given that $F(z)$ converges for all $2 < |z| < 3$.

The numerator is of the same degree in z as the denominator; hence the latter is divided into the former, giving

$$F(z) = -2 + \frac{2z^3 - 6z^2 + 27z - 54}{z^4 - 8z^3 + 21z^2 - 18z}$$

The denominator is found to have simple zeros at $z = 0$ and $z = 2$, and a double zero at $z = 3$. The form of the partial-fraction expansion thus will be

$$F(z) = -2 + \frac{c_1}{z} + \frac{c_2}{z-2} + \frac{c_3}{z-3} + \frac{c_4 z}{(z-3)^2}$$

Combining these terms, equating corresponding coefficients of powers of z, and then solving for the constants yields

$$F(z) = -2 + \frac{3}{z} - \frac{4}{z-2} + \frac{3z}{(z-3)^2}$$

(The constant c_3 is found to equal 0.)

Since $F(z)$ converges for some $|z| > 2$, the term $4/(z-2)$ will yield the time sequence $4(2)^{k-1}$ for $k \geq 1$. Similarly, since $F(z)$ converges for some $|z| < 3$, the term $3z/(z-3)^2$ yields the time sequence $-3k(3)^{k-1}$ for $k \leq 0$. Thus

$$f(k) = -2\,\big|_{k=0} + 3\,\big|_{k=1} - 4(2)^{k-1}\,\big|_{k\geq 1} - 3k(3)^{k-1}\,\big|_{k\leq 0}$$

and, after simplification,

$$f(k) = 3\,\big|_{k=1} - 2^{k+1}\,\big|_{k\geq 0} - k3^k\,\big|_{k\leq 0}.$$

3.8 THE INVERSION INTEGRAL

Consider the generating function

$$F(z) = \sum_{k=-\infty}^{\infty} f(k)z^{-k} \tag{3.51}$$

in the range of z for which the series converges. A circular contour Γ, centered at the origin, may be selected which lies wholly within the region of convergence of $F(z)$. Multiplying both sides of (3.51) by z^{i-1} and integrating over Γ with respect to z yields

$$\oint_\Gamma z^{i-1}F(z)\,dz = \sum_{k=-\infty}^{\infty} \oint_\Gamma f(k)z^{i-k-1}\,dz \tag{3.52}$$

Since the series converges uniformly, it is permissible to interchange the order of integration and summation.

From a basic lemma of complex-variable theory,

$$\oint_C z^\alpha\, dz = 2\pi j \quad \text{for } \alpha = -1$$
$$= 0 \quad \text{for } \alpha \neq -1 \tag{3.53}$$

where C is a contour enclosing the origin. Hence all the terms on the right in (3.52) vanish except the term for which $k = i$. We thus obtain the following *inversion integral*:

$$\boxed{f(k) = \frac{1}{2\pi j}\oint_\Gamma z^{k-1}F(z)\, dz} \tag{3.54}$$

The integral (3.54) is evaluated by determining the residues of the integrand at the poles of $z^{k-1}F(z)$. Observe that as $k \to \infty$, the integrand has an essential singularity at $z = \infty$; similarly, as $k \to -\infty$, there is an essential singularity at $z = 0$. Accordingly it is desirable to evaluate the integral in two parts, one for all $k \geq 0$ and one for $k < 0$; that is

$$f(k) = \frac{1}{2\pi j}\oint_\Gamma z^{k-1}F(z)\, dz\Bigg|_{k\geq 0} + \frac{1}{2\pi j}\oint_\Gamma z^{k-1}F(z)\, dz\Bigg|_{k<0} \tag{3.55}$$

The first integral will contain at most a finite-order pole at $z = 0$, while the second will contain at most a finite-order pole at $z = \infty$, provided, of course, that $F(z)$ does not have an essential singularity at either $z = 0$ or $z = \infty$. Note that for generating functions whose determining functions are zero for $k < 0$ (i.e., for "z transforms"), the second integral must always vanish.

The value of the first integral in (3.55) is given by the sum of the residues of the integrand at the poles of $z^{k-1}F(z)$ contained within Γ. If Γ does not contain any poles, the value of the integral will be zero, and hence $f(k) = 0$ for all $k \geq 0$. Thus

$$f(k)\big|_{k\geq 0} = \sum \text{residues of } z^{k-1}F(z) \text{ at poles of } z^{k-1}F(z) \text{ within } \Gamma \text{ for } k \geq 0 \tag{3.56}$$

For a pole at $z = a$ of multiplicity ν, the residue of $z^{k-1}F(z)$ is given by[12]

$$\boxed{\begin{array}{l}\text{Residue of } z^{k-1}F(z) \\ \text{at } \nu\text{th order pole } z = a\end{array} = \lim_{z\to a}\frac{1}{(\nu-1)!}\frac{d^{\nu-1}}{dz^{\nu-1}}(z-a)^\nu z^{k-1}F(z)} \tag{3.57}$$

[12] See, for example, E. J. Scott, *Transform Calculus with an Introduction to Complex Variables*, Harper and Brothers, New York, 1955, p. 32.

Occasionally in determining the residue at a pole we encounter a pole that is *nonfactorable*. By L'Hospital's rule,

$$\underset{\text{at } z=a}{\text{Residue of}}\ \frac{A(z)}{B(z)} = \lim_{z\to a}(z-a)\frac{A(z)}{B(z)} = \lim_{z\to a}\frac{(z-a)A'(z)+A(z)}{B'(z)}$$

$$\boxed{\text{Residue at nonfactorable pole} = \lim_{z\to a}\frac{A(z)}{B'(z)}} \tag{3.58}$$

In many cases $F(z)$ will have a simple zero at $z = 0$. The integrand $z^{k-1}F(z)$ for $k \geq 0$ will then be analytic at $z = 0$. However, if $F(z)$ contains a νth-order pole at $z = 0$, the integrand will have a pole of order $\nu + 1 - k$ at $z = 0$. Since the order of the pole is a function of k, the residue at $z = 0$ in (3.56) will have to be evaluated separately for each k, from $k = 0$ to $k = \nu + 1$.

If, for $k = 0$, $z^{k-1}F(z)$ has a pole of high order at $z = 0$, it may be preferable to isolate the pole by long division rather than to evaluate (3.56) for various values of k. One need merely divide the denominator of $F(z)$ into the numerator, *beginning with the low-order powers of* z, until the ratio of remainder polynomial to denominator contains a simple zero at $z = 0$. The resulting quotient terms represent the isolated point values. Alternately, we may make use of (3.43) and note that multiplication of a z transform by $z^{-\nu}$ corresponds to a delay of the time sequence by an amount ν; that is, if

$$F(z) = \mathcal{Z}[f(k)] \quad \text{for } k \geq 0$$

then

$$z^{-\nu}F(z) = \mathcal{Z}[f(k-\nu)] \quad \text{for } k \geq \nu$$

The value of the second integral in (3.55) is obtained by utilizing the closed contour $C = C_1 + C_2 + C_3 + C_4$ (Fig. 3-3). Let $\Gamma = -C_3$ and let the radius of C_1 become infinitely large. The contour C will then encircle all poles of $z^{k-1}F(z)$ that lie *outside* the contour Γ.

Since the integrals over C_2 and C_4 cancel each other,

$$\frac{1}{2\pi j}\oint_C z^{k-1}F(z)\,dz = \frac{1}{2\pi j}\int_{C_1} z^{k-1}F(z)\,dz - \frac{1}{2\pi j}\int_{\Gamma} z^{k-1}F(z)\,dz \tag{3.59}$$

The integral over C_1 in (3.59) will vanish provided

$$z^{k-1}F(z) \sim \frac{1}{z^{1+\varepsilon}} \tag{3.60}$$

as $z \to \infty$, where ε is a small positive number. If $F(z)$ is a rational function in z,

$$F(z) = \frac{b_p z^p + b_{p-1} z^{p-1} + \cdots + b_0}{d_q z^q + d_{q-1} z^{q-1} + \cdots + d_0} \tag{3.61}$$

then the integral over C_1 will vanish for all

$$k \leq q - p - 1. \tag{3.62}$$

Thus if $F(z)$ is a rational function in z and the degree in z of the numerator is at least one less than the degree in z of the denominator, then from (3.59) *for all* $k \leq 0$

$$\frac{1}{2\pi j} \oint_{\Gamma} z^{k-1} F(z)\, dz = -\frac{1}{2\pi j} \oint_{C} z^{k-1} F(z)\, dz \tag{3.63}$$

$$= -\sum \text{residues of } z^{k-1}F(z) \text{ at poles of } F(z) \text{ exterior to } \Gamma \tag{3.64}$$

If in (3.61) $p \geq q$, and the integral (3.63) is to be evaluated for all $k \leq 0$, then the denominator must be divided into the numerator until a

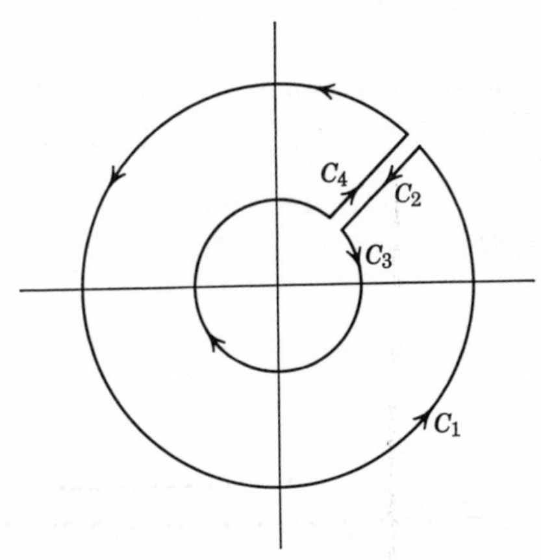

Figure 3-3 Contour for evaluation of inversion integral for $k < 0$.

remainder polynomial is obtained which is of degree one less than the degree of the denominator.[13] The resulting quotient polynomial will be of form $\sum_{l=0}^{p-q} c_l z^l$ and can be inverted by inspection.

The foregoing will be illustrated by applying it to the example given in Section 3.7.

[13] In (3.55) it is necessary to evaluate the second integral only for $k < 0$. In that case it is sufficient if the numerator and denominator of $F(z)$ are of the *same* degree; that is, $p = q$.

Example

$$F(z) = \frac{-2z^4 + 18z^3 - 48z^2 + 63z - 54}{z(z-2)(z-3)^2} \quad \text{for } 2 < |z| < 3$$

The contour Γ will be selected so as to lie in the annulus of convergence. For $k \geq 0$ the residue of

$$z^{k-1}F(z) = \frac{(-2z^4 + 18z^3 - 48z^2 + 63z - 54)z^k}{z^2(z-2)(z-3)^2}$$

must be evaluated at the simple pole $z = 2$. In addition, for $k = 0$ the residue must be evaluated at the double pole $z = 0$, and for $k = 1$ it must be evaluated at the simple pole $z = 0$.

$$\begin{aligned} f(k)\big|_{k\geq 0} &= \text{res } z^{k-1}F(z) \text{ at } z = 2\,\big|_{k\geq 0} \\ &\quad + \text{res } z^{k-1}F(z) \text{ at } z = 0\,\big|_{k=0 \text{ and } k=1} \\ &= -2^{k+1}\big|_{k\geq 0} + 0\,\big|_{k=0} + 3\,\big|_{k=1} \end{aligned}$$

The numerator and denominator of $F(z)$ are of the same degree in z. Hence for $k < 0$, $f(k)$ is given simply by the negative of the residue at the double pole $z = 3$:

$$\begin{aligned} f(k)\big|_{k<0} &= -\text{res } z^{k-1}F(z) \quad \text{at } z = 3 \\ &= -\frac{d}{dz}(z-3)^2 z^{k-1}F(z)\,\big|_{z=3} \\ &= -k3^k \end{aligned}$$

Thus for *all* k,

$$f(k) = 3\,\big|_{k=1} - 2^{k+1}\big|_{k\geq 0} - k3^k\big|_{k<0}$$

3.9 INITIAL-VALUE DETERMINATION FOR A z TRANSFORM

Given the z transform $F(z)$ of the time sequence $\{f(k)\}$, $f(k) = 0$ for $k < 0$, it is possible to determine the initial value $f(0)$ without performing an inverse transformation on $F(z)$, provided $f(0)$ is finite. Since $f(k) = 0$ for $k < 0$, the expression

$$F(z) = f(0) + f(1)z^{-1} + f(2)z^{-2} + \cdots \rightarrow \infty$$

converges uniformly for all $|z| > R_c$, including $|z| = \infty$. Note that as $z \to \infty$, $F(z) \to f(0)$. Hence

$$\boxed{f(0) = \lim_{z \to \infty} F(z)} \tag{3.65}$$

3.10 INTERMEDIATE-VALUE DETERMINATION FOR A $\mathcal{Z}$ TRANSFORM

If we are given an $F(z)$ which converges for all $|z| > R_c$, we may determine $f(k)$ for any particular k as follows:

$$F(z) = f(0) + f(1)\, z^{-1} + f(2)z^{-2} + \cdots \to \infty$$

Differentiating with respect to z and multiplying by $-z^2$ yields

$$-z^2 \frac{d}{dz} F(z) = f(1) + 2f(2)z^{-1} + 3f(3)z^{-2} + \cdots$$

which also converges uniformly for all $|z| > R_c$. Hence

$$f(1) = \lim_{z \to \infty} \left[-z^2 \frac{d}{dz} F(z) \right] \tag{3.66}$$

and, more generally,

$$\boxed{f(k) = \lim_{z \to \infty} \frac{1}{k!} \left(-z^2 \frac{d}{dz} \right)^k F(z) \quad \text{for any } k \geq 0} \tag{3.67}$$

3.11 FINAL-VALUE DETERMINATION FOR A $\mathcal{Z}$ TRANSFORM

If, as k goes to infinity, $f(k)$ approaches a finite limit, the value of this limit can be found as follows. Consider an $F(z)$ which converges for all $|z| > R_c$. Its partial-fraction expansion will consist of a series of terms having the forms given in the left-hand column of Table 3-1. Reference to column I of Table 3-1 then shows that each term of this expansion leads to a component time sequence of form $k^j a_i^{k-1}$ for $k \geq 1$. If $f(k)$ is to approach a finite limit as k goes to infinity, then for all $j > 0$, it is necessary that

$$|a_i| < 1$$

The limit of all components of $f(k)$ for which $j > 0$ will thus be zero. Only if $j = 0$ and $a_i = 1$ can a time sequence be obtained which has a finite nonzero limit; that is, if $\mathcal{Z}[f(k)] = c/(z - 1)$, then $\lim_{k\to\infty} f(k) = c$.

Observe that if the terms in the partial-fraction expansion of $F(z)$ are all multiplied by $z - 1$ and then $z \to 1$, all terms will vanish except the term for which $j = 0$ and $a = 1$ (if such a term is present). The value of the latter is the desired "final" value of $f(k)$; that is,

$$\boxed{\lim_{k\to\infty} f(k) = \lim_{z\to 1} (z - 1)F(z)} \tag{3.68}$$

provided the limit of $f(k)$ exists!

3.12 EXAMPLE OF A PROBLEM IN PROBABILITY THEORY

We shall now give an example which illustrates the use of the transformation calculus for partial difference equations.

Consider an unbalanced coin which is tossed k times. Let the probability of HEADS on any single toss be h. It is desired to determine the probability $P(l, k)$ that exactly l HEADS will be obtained in k tosses.

To solve this problem we shall make use of the fact that to obtain exactly l HEADS after k tosses, we must either already have l HEADS after $k - 1$ tosses and then obtain a TAIL on the kth toss, or we must have $l - 1$ HEADS after $k - 1$ tosses and then obtain a HEAD on the kth toss.

Let us denote the probability of obtaining TAILS by t; that is, $t = 1 - h$. Then for any $l \leq k$ the process is described by the following linear partial difference equation in two independent variables:

$$P(l, k) = hP(l - 1, k - 1) + tP(l, k - 1) \tag{3.69}$$

Let $M(z, k)$ be the generating function of $P(l, k)$ with respect to the variable l; that is,

$$M(z, k) = \sum_{l=0}^{k} P(l, k)z^{-l} \tag{3.70}$$

Then, from (3.43),

$$\sum_{l=0}^{k} P(l - 1, k)z^{-l} = z^{-1}M(z, k) \tag{3.71}$$

The difference equation can now be transformed with respect to l:

$$\begin{aligned} M(z, k) &= hz^{-1}M(z, k - 1) + tM(z, k - 1) \\ &= (t + hz^{-1})M(z, k - 1) \end{aligned} \tag{3.72}$$

Now let $N(z, \zeta)$ be the generating function of $M(z, k)$ with respect to k. Then

$$N(z, \zeta) = \zeta^{-1}(t + hz^{-1})N(z, \zeta)$$

or

$$[1 - \zeta^{-1}(t + hz^{-1})]N(z, \zeta) = 0 \tag{3.73}$$

The last equation is the generating function of a first-order linear homogeneous difference equation. Hence the inverse transformation of (3.73) with respect to ζ yields

$$M(z, k) = c(t + hz^{-1})^k \tag{3.74}$$

where c is an arbitrary constant. Expanding the right side of (3.74), we have

$$\begin{aligned} M(z, k) &= c\left[t^k + kt^{k-1}hz^{-1} + \frac{k(k-1)}{2!}t^{k-2}h^2z^{-2} + \cdots + h^kz^{-k}\right] \\ &= c\sum_{l=0}^{k}\binom{k}{l}t^{k-l}h^lz^{-l} \end{aligned} \tag{3.75}$$

Comparing this last result with the definition of $M(z, n)$ in (3.70) shows that

$$P(l, k) = c\binom{k}{l}t^{k-l}h^l \tag{3.76}$$

Now for $k = l = 1$, $P(1, 1) = h$. Hence $c = 1$ and the desired solution is

$$P(l, k) = \binom{k}{l}t^{k-l}h^l \tag{3.77}$$

Equation (3.77) is known as Bernoulli's theorem. Note that $P(l, k)$ represents the lth term in the binomial expansion of $(t + h)^k$. Accordingly, $\{P(l, k)\}$ (for a fixed k) is also known as the *binomial distribution.*

3.13 $\mathcal{Z}$ TRANSFORM OF THE PRODUCT $f(k)\,h(k)$

We shall now derive a formula for the z transform of the *product* of two time sequences, given the z transforms of the individual sequences.

Consider two sequences $\{f(k)\}$ and $\{h(k)\}$, where $f(k) = h(k) = 0$ for $k < 0$, having the z transforms:

$$\begin{aligned} F(z) &= \mathcal{Z}[f(k)] \quad \text{for } |z| > R_f \\ H(z) &= \mathcal{Z}[h(k)] \quad \text{for } |z| > R_h \end{aligned} \tag{3.78}$$

R_f and R_h are the two radii of absolute convergence.

The z transform of the product $f(k)h(k)$ may be expressed as

$$\mathcal{Z}[f(k)h(k)] = \sum_{k=0}^{\infty} f(k)h(k)z^{-k} \tag{3.79}$$

for $|z| > R_c$. Now from (3.54)

$$h(k) = \frac{1}{2\pi j} \oint_\Gamma \zeta^{k-1} H(\zeta)\, d\zeta \tag{3.80}$$

Upon substituting (3.80) in (3.79),

$$\mathcal{Z}[f(k)h(k)] = \frac{1}{2\pi j} \sum_{k=0}^{\infty} \oint_\Gamma f(k)H(\zeta)\zeta^{k-1}z^{-k}\, d\zeta \tag{3.81}$$

Since (3.79) converges uniformly for $|z| > R_c$, we may interchange the order of summation and integration. Thus

$$\mathcal{Z}[f(k)h(k)] = \frac{1}{2\pi j} \oint_\Gamma \zeta^{-1}H(\zeta) \sum_{k=0}^{\infty} f(k)(\zeta^{-1}z)^{-k}\, d\zeta \tag{3.82}$$

Since $F(z)$ is the z transform of $\{f(k)\}$,

$$\sum_{k=0}^{\infty} f(k)(\zeta^{-1}z)^{-k} = F(\zeta^{-1}z)$$

and hence

$$\boxed{\mathcal{Z}[f(k)h(k)] = \frac{1}{2\pi j} \oint_\Gamma \zeta^{-1}H(\zeta)F(\zeta^{-1}z)\, d\zeta} \tag{3.83}$$

The magnitude of z must be chosen to provide an annular region of convergence between the singularities of $\zeta^{-1}H(\zeta)$ and those of $F(\zeta^{-1}z)$. The contour Γ is then taken in this region, that is, where

$$R_h < |\zeta| < |z|/R_f \tag{3.84}$$

3.14 PARSEVAL'S THEOREM

If the sequences $\{f(k)\}$ and $\{h(k)\}$ in (3.78) are such that (3.84) is satisfied for $|z| = 1$, we may let $|z| = 1$ in (3.83) and write

$$\sum_{k=0}^{\infty} f(k)h(k) = \frac{1}{2\pi j} \oint_\Gamma \zeta^{-1}H(\zeta)F(\zeta^{-1})\, d\zeta \tag{3.85}$$

If we now let $f(k) = h(k)$, and provided $R_f < 1$, we obtain the discrete-time form of *Parseval's theorem*,

$$\boxed{\sum_{k=0}^{\infty} [f(k)]^2 = \frac{1}{2\pi j} \oint_\Gamma z^{-1}F(z)F(z^{-1})\, dz} \tag{3.86}$$

where the more familiar z has been substituted for ζ. If we make use of (3.54), we may express (3.86) in the compact form

$$\boxed{\sum_{k=0}^{\infty}[f(k)]^2 = \mathcal{G}^{-1}[F(z)F(z^{-1})]\Big|_{k=0}} \tag{3.87}$$

The requirement that $R_f < 1$ implies that $f(k) \to 0$ as $k \to \infty$. The theorem provides a convenient means for obtaining the sum of the squares of a convergent sequence.

Example

Let

$$\begin{aligned} f(k) &= a^k, \qquad k \geq 0 \\ &= 0, \qquad k < 0 \\ &|a| < 1 \end{aligned}$$

Then

$$F(z) = \frac{1}{1 - az^{-1}} \quad \text{for } |z| > |a|$$

$$F(z)F(z^{-1}) = \frac{-z}{a(z-a)(z-1/a)}, \qquad \frac{1}{|a|} > |z| > |a|$$

From (3.86),

$$\sum_{k=0}^{\infty}[f(k)]^2 = -\frac{1}{2\pi j}\oint_{\Gamma} \frac{dz}{a(z-a)(z-1/a)}$$

where Γ may be taken as the circle of unit radius in the z plane. Evaluating the residue at the enclosed pole $z = a$,

$$\sum_{k=0}^{\infty}[f(k)]^2 = \frac{1}{1-a^2}$$

3.15 RELATION TO THE LAPLACE TRANSFORM

We shall now examine the connection between the Laplace transform for continuous-time functions and the z transformation.

If a function $\{f(t)\}$ is both sectionally continuous and of exponential order, that is, if as $t \to \infty$, we can find a finite number c such that $|f(t)|\, e^{-ct}$ is bounded, then the integral

$$F(s) = \int_{0-}^{\infty} f(t)e^{-st}\, dt \tag{3.88}$$

exists for all Re $s > \sigma_a$, where σ_a is the least value of c which assures the boundedness of $|f(t)|\,e^{-ct}$.[14] The integral (3.88) is known as the *Laplace transform*[15] of $\{f(t)\}$, $\mathscr{L}[f(t)]$. The number σ_a is the abscissa of absolute convergence associated with $\{f(t)\}$ in (3.88).

It is of interest to investigate how the Laplace transform can be applied to the solution of discrete-time problems. Since neither the integral nor the derivative of a sequence is defined, the Laplace transform of a sequence clearly does not exist. However, this difficulty can be circumvented by introducing a *carrier function.* The latter is a function that is Laplace transformable and that can be made to "carry" a given sequence.

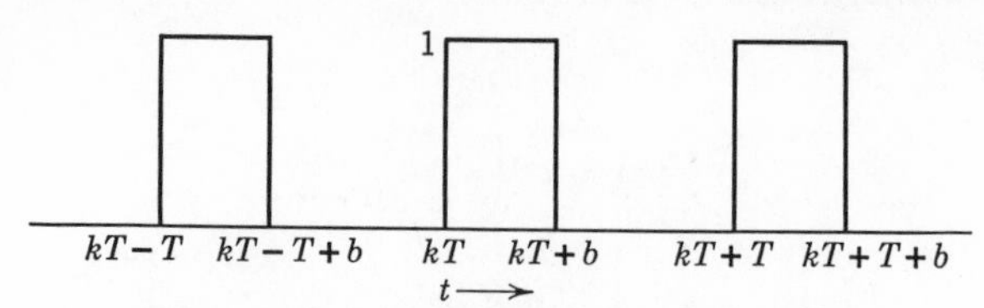

Figure 3-4 Series of flat pulses of duration b.

Consider the function $\{p(t;\,T,\,b)\}$, which is defined by

$$\begin{aligned} p(t;\,T,\,b) &= 1 \quad \text{for } kT \leq t < kT + b \\ &= 0 \quad \text{elsewhere} \end{aligned} \tag{3.89}$$

where $T > 0$, $0 < b \leq T$, and $k = 0, 1, 2, \ldots$. This function represents a series of flat pulses of duration b, as shown in Fig. 3-4. Its Laplace transform is given by

$$\mathscr{L}[p(t;\,T,\,b)] = \frac{1 - e^{-sb}}{s}\sum_{k=0}^{\infty} e^{-kTs}, \qquad \operatorname{Re} s > 0 \tag{3.90}$$

Now consider a sequence

$$\{f(k)\} = f(0),\, f(1),\, f(2), \ldots \tag{3.91}$$

If we arrange to have the amplitudes of the pulses of $\{p(t;\,T,\,b)\}$ bear a one-to-one correspondence to the values $f(k)$, then the *modulated* pulse series $\{f(t;\,T,\,b)\}$ is obtained for which the Laplace transform is

$$\mathscr{L}[f(t;\,T,\,b)] = \frac{1 - e^{-sb}}{s}\sum_{k=0}^{\infty} f(k)e^{-kTs} \tag{3.92}$$

This function is illustrated in Fig. 3-5.

[14] This is a sufficient condition. For some functions σ_a may be less than the least value of c.

[15] For a detailed description of the Laplace transform, see G. Doetsch, *Theorie und Anwendung der Laplace-Transformation*, Springer, Berlin, 1937; D. Widder, *The Laplace Transform*, Princeton University Press, Princeton, N.J., 1941; E. Scott, *Transform Calculus*, Harper and Brothers. New York, 1955; or R. V. Churchill, *Operational Mathematics* (2nd ed.), McGraw-Hill Book Co., New York, 1958.

A linear discrete-time system for which an input $\{u(k)\}$ yields an output $\{y(k)\}$ (where both are expressed in the form of modulated pulse series) is characterized by the ratio of the Laplace transforms of output and input that is, from (3.92),

$$H(s) = \frac{\sum_{k=0}^{\infty} y(k)e^{-kTs}}{\sum_{k=0}^{\infty} u(k)e^{-kTs}} \tag{3.93}$$

This *transfer function* $H(s)$ is rational in e^{sT}.

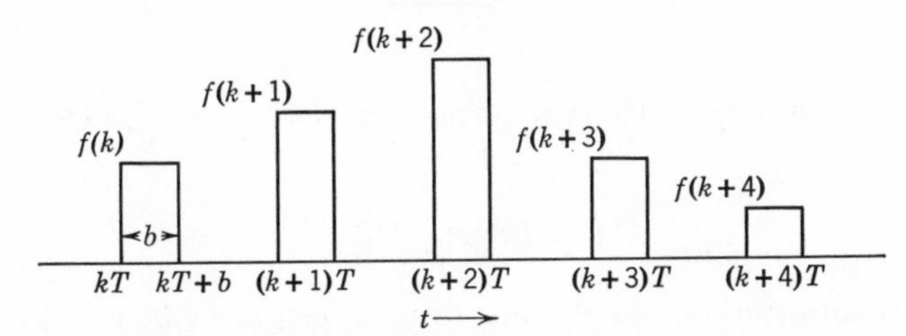

Figure 3-5 Modulated pulse series $\{f(t; T, b)\}$.

A comparison of (3.92) with (3.23) shows that we may write

$$\mathscr{L}[f(t; T, b)] = \frac{1 - e^{-sb}}{s} \mathcal{Z}[f(k)]\Big|_{z=e^{sT}} \tag{3.94}$$

The Laplace transform of any modulated pulse function can thus be readily obtained from the z transforms derived in Section 3.5.

If a pulse series such as (3.92) is supplied as an input to a linear discrete-time system of transfer function $H(s)$, the output sequence will be given by[16]

$$y(k) = \mathcal{Z}^{-1}\left[\frac{s}{1 - e^{-sb}} Y(s; T, b)\Big|_{e^{sT}=z}\right] \tag{3.95}$$

where

$$Y(s; T, b) = H(s)\mathscr{L}[f(t; T, b)] \tag{3.96}$$

Thus by introducing the Laplace-transformable carrier function (3.90), and utilizing relations (3.93), (3.94), and (3.95), the Laplace transform can also be used for the solution of discrete-time system problems.

In (3.92) the pulse width b may be of any value,

$$0 < b \leq T$$

If $b = T$, the gaps between the pulses in Fig. 3-5 vanish and the pulse series takes on the form of a *staircase function*. The use of modulated

[16] The symbol $\mathcal{Z}^{-1}$ denotes "inverse z transformation of."

staircase functions for the Laplace transform solution of discrete-time problems has been employed by many writers.[17] The term *jump function* is also frequently used for these functions.

The pulse width b may be permitted to approach zero provided the amplitude is increased so as to keep the integral of the pulse finite and equal to unity. The resulting "pulse" is referred to as an *impulse* and the modulation is with respect to its integral (i.e., area) rather than its amplitude. The Laplace transform of a pulse of width b and amplitude $1/b$ as b goes to zero is defined to be

$$\lim_{b \to 0} \frac{1}{b} \frac{1 - e^{-sb}}{s} = 1 \tag{3.97}$$

Thus if an impulse series is used as a carrier function, (3.92) becomes simply

$$\mathscr{L}[f(t; T)] = \sum_{k=0}^{\infty} f(k) e^{-kTs} \tag{3.98}$$

With the substitution $z = e^{sT}$, this form is seen to be identical to that of the z transform of $\{f(k)\}$. There is, however, an important distinction between (3.98) and (3.23). The z transform represents a transformation of a sequence. In contrast, $\mathscr{L}[f(t; T)]$ as given in (3.98) represents the Laplace transform of a function defined over the continuous time domain; this function is implied to be zero everywhere except at $t = kT$ where it consists of an impulse of *area value* $f(k)$.

The concept of impulse modulation was introduced by Linvill.[18] It has considerable intuitive appeal from an engineering point of view. This is particularly true for system problems in which some variables vary continuously and others vary discretely. Although this applies also to the modulated finite pulse series (including the staircase), the latter tend to lead to somewhat more complicated expressions.

One drawback of the concept of impulse modulation as previously defined is that the definition is highly questionable from a mathematical point of view. This drawback can be overcome by approaching the formulation in a different way. Unfortunately, the alternate derivation is devoid of the physical insight afforded by viewing an impulse as the area of a narrow but large-amplitude pulse. It does serve, however, as a basis for justifying the impulse modulation method.

[17] See in particular M. F. Gardner and J. L. Barnes, *Transients in Linear Systems*, vol. 1, John Wiley and Sons, New York, 1942, pp. 286–331; and E. Scott, *op. cit.*, pp. 162–196.

[18] W. K. Linvill, "Sampled-Data Control Systems Studied Through Comparison of Sampling with Amplitude Modulation," *Trans. AIEE*, vol. 70, part 2, pp. 1779–1788, 1951.

Let $\{f(t)\}$ be a continuous function in a finite interval $(a < \tau < b)$ and let it vanish everywhere outside this interval. Then we can define a symbolic function $\delta(t)$ (also referred to as the Dirac delta function) by means of the integral

$$\int_{-\infty}^{\infty} \delta(t - \tau) f(t)\, dt = f(\tau) \tag{3.99}$$

Thus if a continuous function $\{f(t)\}$ is multiplied by the delta function $\delta(t - \tau)$ and the product integrated over all t, the result is the value of $\{f(t)\}$ at $t = \tau$. The effect is a *sampling* action on $\{f(t)\}$.

It is now possible to establish a basis for modulating an impulse function with a sequence. Given a sequence $\{f(k)\}$, $k \geq 0$, we stipulate a continuous function $\{f_1(t)\}$, $t \geq 0$, such that

$$\{f_1(t)\}\big|_{t=kT} = f_1(kT) = f(k) \tag{3.100}$$

where T is a positive constant and $\{f_1(t)\}$ is otherwise arbitrary. Then for a *particular* k,

$$\int_{-\infty}^{\infty} \delta(t - kT) f_1(t)\, dt = f(k) \tag{3.101}$$

Using $f_1(t)e^{-st}$ instead of $f_1(t)$, we have

$$\int_{-\infty}^{\infty} \delta(t - kT) f_1(t) e^{-st}\, dt = f(k) e^{-kTs} \tag{3.102}$$

which is seen to be the Laplace transform of $\delta(t - kT) f_1(t)$.

If we now introduce the delta-function series

$$\delta(t; T) = \sum_{k=-\infty}^{\infty} \delta(t - kT) \tag{3.103}$$

then

$$\int_{-\infty}^{\infty} \delta(t; T) f_1(t) e^{-st}\, dt = \sum_{k=0}^{\infty} f(k) e^{-kTs} \tag{3.104}$$

This is precisely the result obtained in (3.98) from the so-called process of impulse modulation. We may thus regard the symbolic function $\delta(t; T)$ defined by (3.103) and (3.99) as a *carrier* which is modulated by a continuous function $\{f_1(t)\}$. If desired, $\delta(t)$ may be referred to as a unit impulse *provided it is clearly understood that this is a symbolic function defined only as expressed by* (3.99).[19]

The use of the Laplace transform—either in terms of a modulated pulse series or the delta function series—is most suitable for problems in which some of the variables are actually in the form of short pulses or staircase

[19] B. Friedman, *Principles and Techniques of Applied Mathematics*, John Wiley and Sons, New York, 1956.

functions. No advantage, neither computational nor intuitive, is gained over the generating function in the case of true discrete-time systems. It is of passing interest to point out also that the carrier functions are not restricted to flat pulses or impulses; any Laplace-transformable function which vanishes everywhere outside the interval $kT \leq t < kT + T$ may be used instead of $p(t; T, b)$ in (3.92).

REFERENCES

Barker, R. H., "The Pulse Transfer Function and Its Application to Sampling Servo-Systems," *Proc. IEE*, vol. 99, part 4, pp. 302–317, London, December 1952.

Barker, R. H., *The Theory of Pulse Monitored Servomechanisms and Their Use for Prediction*, Report No. 1046, Signals Research and Development Establishment, Ministry of Supply, Christchurch, Hants, England, November 1950.

Bergen, A. R. and J. R. Ragazzini, "Sampled-Data Processing Techniques for Feedback Control Systems," *Trans. AIEE* vol. 73, part 2, pp. 236–247, November 1954.

Bridgeland, T. F., "A Linear Algebraic Formulation of the Theory of Sampled-Data Control," *J. Soc. Ind. Appl. Math.*, vol. 7, no. 4, pp. 431–446, December 1959.

Brown, R. G. and J. W. Nilsson, *Introduction to Linear System Analysis*, John Wiley and Sons, New York, 1962.

Churchill, Ruel, *Operational Mathematics* (2nd ed.), McGraw-Hill Book Co., New York, 1958.

Freeman, H., "A Comparison of Methods for the Analysis of Pulsed Linear Systems," *Proc. Nat. Electron. Conf.*, vol. 15, pp. 1032–1043, 1959.

Freeman, H. and O. Lowenschuss, "Bibliography of Sampled-Data Control Systems and $\mathcal{Z}$-transform Applications," *IRE Trans. Autom. Control*, vol. PGAC-4, pp. 28–30, March 1958.

Gardner, M. F. and J. L. Barnes, *Transients in Linear Systems*, vol. 1, John Wiley and Sons, New York, 1942.

Helm, H. A., "The $\mathcal{Z}$ Transformation," *Bell System Tech. J.*, vol. 38, no. 1, pp. 177–196, January 1959.

Jordan, Charles, *Calculus of Finite Differences* (2nd ed.), Chelsea Publ. Co., New York, 1950.

Jury, E. I., "Analysis and Synthesis of Sampled-Data Control Systems," *Trans. AIEE*, vol. 73, part 1, pp. 332–346, September 1954.

Jury, E. I., *Sampled-Data Control Systems*, John Wiley and Sons, New York, 1958.

Jury, E. I., *Theory and Application of the $\mathcal{Z}$-Transform Method*, John Wiley and Sons, New York, 1964.

Knopp, K., *Infinite Sequences and Series*, Dover Publications, New York, 1956.

Kuo, Benjamin, *Analysis and Synthesis of Sampled-Data Control Systems*, Prentice-Hall, Englewood Cliffs, N.J., 1963.

Ragazzini, J. R. and G. F. Franklin, *Sampled-Data Control Systems*, McGraw-Hill Book Co., New York, 1958.

Ragazzini, J. R. and L. A. Zadeh, "The Analysis of Sampled-Data Systems," *Trans. AIEE*, vol. 71, part 2, pp. 225–232, 1952.

Scott, E. J., *Transform Calculus with an Introduction to Complex Variables*, Harper and Brothers, New York, 1955.

Stromer, P. R., "A Selective Bibliography on Sampled-Data Systems," *IRE Trans. on Autom. Control*, vol. PGAC-6, pp. 112–114, December 1958.

Thomson, W. T., *Laplace Transformation* (2nd ed.), Prentice-Hall, Englewood Cliffs, N.J., 1960.

Titchmarsh, E. C., *The Theory of Functions* (2nd ed.), Oxford University Press, London, 1939.

Tou, Julius, *Digital and Sampled-Data Control Systems*, McGraw-Hill Book Co., New York, 1959.

Tsypkin, Y. Z., *Theory of Impulse Systems*, State Publisher for Physical-Mathematical Literature, Moscow, 1958.

Widder, D. V., *Advanced Calculus* (2nd ed.), Prentice-Hall, Englewood Cliffs, N.J., 1961.

CHAPTER 4

Sampling of Continuous-Time Functions

4.1 INTRODUCTION

In the study of functions of a discrete variable we recognize two kinds of problems. The first is concerned with the manipulation and solution of equations involving true time sequences. The second is concerned with the representation of functions of a continuous variable by functions of a discrete variable. The preceding chapters have been concerned with the first type; in this chapter we shall consider the second type.

The selecting of a time sequence $\{f(kT)\}$ to represent a continuous-time function $\{f(t)\}$ is known as sampling. In effect one "samples" the continuous-time function at specified instants of time. The result is usually an approximation which improves as the interval between samples is decreased. Except for certain idealized cases to be described, a time sequence cannot fully describe a continuous-time function, and hence the reconstruction of an $\{f(t)\}$ from its samples is always subject to some ambiguity.

4.2 LAPLACE TRANSFORM OF A SAMPLED CONTINUOUS-TIME FUNCTION

We shall now derive the Laplace transform of $\{f(t; T)\}$, that is, of the function obtained by modulating a continuous-time function $\{f(t)\}$ by a delta-function carrier as described in Section 3.15. Two alternate but equivalent forms for this Laplace transform will be obtained. One of these will yield the z transform of the sequence $\{f(kT)\}$ if we make the substitution $z = e^{sT}$. The other leads to an infinite summation in the complex frequency domain and provides a basis for determining the conditions under which a continuous-time function can be perfectly reconstructed from its samples.

Consider a Laplace transformable function $\{f(t)\}$ that vanishes for

$t < 0$. If the function is evaluated for $t = kT$, where $k = 0, 1, 2, \ldots, \infty$, the resulting sequence

$$\{f(kT)\} = f(0),\ f(T),\ f(2T), \ldots \tag{4.1}$$

may be used to modulate the delta-function carrier (3.103). Then, in accordance with (3.98) and (3.104),

$$f(t; T) = \delta(t; T)f(t) = f(t)\sum_{k=0}^{\infty}\delta(t - kT) \tag{4.2}$$

In general, we can express the product of two functions $\{f_1(t)\}$ and $\{f_2(t)\}$ whose respective Laplace transforms $F_1(\lambda)$ and $F_2(p)$ exist as follows:

$$f_1(t)f_2(t) = \left(\frac{1}{2\pi j}\right)^2 \int_{c_1-j\infty}^{c_1+j\infty} d\lambda \int_{c_2-j\infty}^{c_2+j\infty} dp\, e^{t(\lambda+p)}F_1(\lambda)F_2(p) \tag{4.3}$$

where $c_1 > \sigma_{a1}$ and $c_2 > \sigma_{a2}$, and σ_{a1} and σ_{a2} are the abscissae of absolute convergence for $F_1(\lambda)$ and $F_2(p)$, respectively. Substituting $s = \lambda + p$ and rearranging yields

$$f_1(t)f_2(t) = \frac{1}{2\pi j}\int_{c-j\infty}^{c+j\infty} ds\, e^{Ts}\left[\frac{1}{2\pi j}\int_{c-j\infty}^{c+j\infty} d\lambda\, F_1(\lambda)F_2(s - \lambda)\right] \tag{4.4}$$

The path of integration is taken in the analytic strip where c satisfies the condition $\sigma_{a1} < c < \sigma - \sigma_{a2}$, and $\sigma > \max(\sigma_{a1}, \sigma_{a2}, \sigma_{a1} + \sigma_{a2})$. The term in brackets in (4.4) is recognized as the Laplace transform of $\{f_1(t)f_2(t)\}$; that is,

$$\mathscr{L}[f_1(t)f_2(t)] = \frac{1}{2\pi j}\int_{c-j\infty}^{c+j\infty} d\lambda\, F_1(\lambda)F_2(s - \lambda) \tag{4.5}$$

The right-hand side of (4.5) is known as the complex convolution integral. The selection of σ is such that the poles of $F_1(\lambda)$ will lie to the left of $\lambda = \sigma_{a1}$ and those of $F_2(s - \lambda)$ will lie to the right of $\lambda = \sigma - \sigma_{a2}$.

Equation (4.5) will now be used to determine the Laplace transform of $\{f(t; T)\}$ as given by (4.2). Let $F_1(\lambda) = F(\lambda)$ be the Laplace transform of $\{f(t)\}$. It will be assumed that $F(\lambda)$ is a meromorphic function.[1] From (3.102),

$$\int_{0-}^{\infty} \delta(t - kT)e^{-st}\, dt = e^{-sTk} \tag{4.6}$$

Hence the Laplace transform of the summation of (4.2) is given by

$$\mathscr{L}\left[\sum_{k=0}^{\infty}\delta(t - kT)\right] = \sum_{k=0}^{\infty} e^{-sTk} = \frac{1}{1 - e^{-sT}} \tag{4.7}$$

[1] A meromorphic function is a function all of whose singularities are poles.

for $|e^{sT}| > 1$. Letting $F_2(s - \lambda) = \dfrac{1}{1 - e^{T(\lambda - s)}}$ and substituting in (4.5) yields

$$\boxed{\mathscr{L}[f(t; T)] = \frac{1}{2\pi j}\int_{c-j\infty}^{c+j\infty} d\lambda \, \frac{F(\lambda)}{1 - e^{T(\lambda - s)}}} \tag{4.8}$$

The integral (4.8) can be evaluated by the method of residues upon forming a closed, semicircular contour with a radius which approaches infinity over either the left-half plane or the right-half plane. Thus

$$\mathscr{L}[f(t; T)] = \frac{1}{2\pi j}\oint_{\mathrm{I}+\uparrow} \frac{F(\lambda)\, d\lambda}{1 - e^{T(\lambda - s)}} - \frac{1}{2\pi j}\int_{\mathrm{I}} \frac{F(\lambda)\, d\lambda}{1 - e^{T(\lambda - s)}} \tag{4.9}$$

$$= \frac{1}{2\pi j}\oint_{\mathrm{II}+\uparrow} \frac{F(\lambda)\, d\lambda}{1 - e^{T(\lambda - s)}} - \frac{1}{2\pi j}\int_{\mathrm{II}} \frac{F(\lambda)\, d\lambda}{1 - e^{T(\lambda - s)}} \tag{4.10}$$

where I and II are the semicircular paths and $\uparrow$ represents the path of integration from $c - j\xi$ to $c + j\xi$ as $\xi \to \infty$. The complete contours are shown in Fig. 4-1. It is seen that the contour I + $\uparrow$ encloses the poles of $F(\lambda)$ and that the contour II + $\uparrow$ encloses the poles of $1/(1 - e^{T(\lambda - s)})$.

Let us consider first the expression given by (4.9). For the integral over the closed contour I + $\uparrow$ we obtain in accordance with Cauchy's residue theorem

$$\frac{1}{2\pi j}\oint_{\mathrm{I}+\uparrow} \frac{F(\lambda)\, d\lambda}{1 - e^{T(\lambda - s)}} = \sum \text{residues of } \frac{F(\lambda)}{1 - e^{T(\lambda - s)}} \text{ at poles of } F(\lambda) \tag{4.11}$$

The integral over the semicircular path I is determined as follows. If $F(\lambda)$ is a rational function with real coefficients, then as $R \to \infty$, $F(\lambda)$ will approach B/λ^ν, where ν is an integer and B is a real number. Let $\lambda = Re^{j\theta}$. Then $d\lambda = jRe^{j\theta}\, d\theta$, and

$$\frac{1}{2\pi j}\int_{\mathrm{I}} \frac{F(\lambda)\, d\lambda}{1 - e^{T(\lambda - s)}} = \lim_{R \to \infty} \frac{1}{2\pi j}\int_{\pi/2}^{3\pi/2} \frac{jB\, d\theta}{R^{\nu - 1} e^{j\theta(\nu - 1)}(1 - e^{TR\cos\theta} e^{jTR\sin\theta} e^{-Ts})} \tag{4.12}$$

$$\begin{aligned} &= 0 && \text{for } \nu > 1 \\ &= B/2 && \text{for } \nu = 1 \\ &= \text{undefined} && \text{for } \nu \leq 0 \end{aligned} \tag{4.13}$$

It is apparent from the initial-value theorem for Laplace transforms[2] that

[2] M. F. Gardner and J. L. Barnes, *Transients in Linear Systems*, vol. 1, John Wiley and Sons, New York, 1942, pp. 267–268.

$B = f(0_+)$ and that, of course, $f(0_+) = 0$ for $\nu > 1$. Hence from (4.13) and (4.11), the expression given by (4.9) can be rewritten in the form

$$\mathscr{L}[f(t; T)] = \sum \underset{\text{at poles of } F(\lambda)}{\text{residues of}} \frac{F(\lambda)}{1 - e^{T(\lambda - s)}} - \frac{f(0_+)}{2} \tag{4.14}$$

In accordance with (4.13) it is clear that if $F(\lambda)$ is rational, the degree of the denominator must exceed that of the numerator by at least unity; that is, for the Laplace transform (4.14) to exist it is necessary that $\nu \geq 1$.

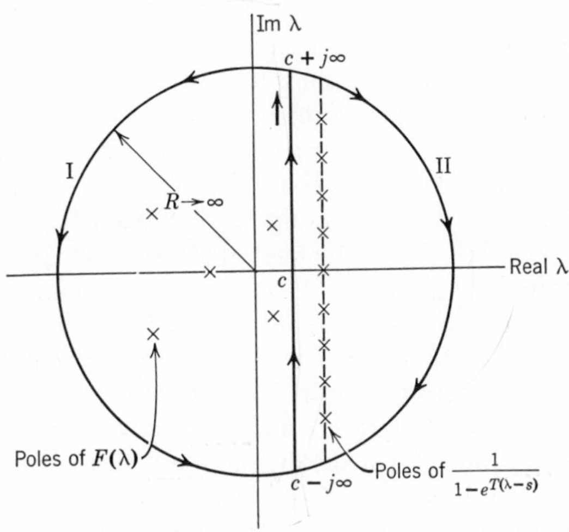

Figure 4-1 Location of poles and paths of integration for evaluation of $\mathscr{L}[f(t; T)]$. (*Note*: poles of $F(\lambda)$ lie to left of $\lambda = c$.)

It follows also from the foregoing that if $F(\lambda)$ is meromorphic but not rational, the integral over the path I will vanish if as $R \to \infty$, $F(\lambda)$ approaches zero more rapidly than as $1/\lambda$. The Laplace transform $\mathscr{L}[f(t; T)]$ will then be given simply by (4.14) with the term $-f(0_+)/2$ deleted.

Observe that the integral over the path I in (4.9) vanishes for all $F(\lambda)$ except in the case for which $F(\lambda) \to B/\lambda$ as $R \to \infty$.[3] The latter case arises when $\{f(t)\}$ possesses a discontinuity at $t = 0$. The integral over I then takes on the value $f(0_+)/2$ and this value must be subtracted from the sum of the residues obtained from (4.11), as indicated by (4.14). This method of obtaining the Laplace transform of $\{f(kT)\}$ is in accord with the

[3] It is assumed, of course, that the integral is defined.

convention that at a point of discontinuity (in this case the point $t = 0$) $f(t)$ is assumed to be given by the mean of the values to either side of the discontinuity.[4] For example, if $F(\lambda) = 1/(\lambda + a)$, $\nu = 1$, and use of (4.14) leads to

$$\mathscr{L}[f(t; T)] = \frac{1}{1 - e^{-aT}e^{-sT}} - \frac{1}{2}, \qquad \text{Re } s > \sigma_a$$

which upon long-division expansion and term-by-term inversion yields

$$\begin{aligned} f(kT) &= 0 && \text{for } k < 0 \\ &= \tfrac{1}{2} && \text{for } k = 0 \\ &= e^{-akT} && \text{for } k > 0 \end{aligned}$$

It is possible in practice to avoid the need to subtract $f(0_+)/2$ in (4.14) when $\nu = 1$. We need merely shift the discontinuity from the point $t = 0$ to a point $t = 0_-$. Thus instead of $F(\lambda) = 1/(\lambda + a)$, we write (or at least *imply*) $F(\lambda) = e^{\varepsilon\lambda}/(\lambda + a)$, where ε is a vanishingly small positive number. $F(\lambda)$ is now no longer rational; it is, however, still meromorphic. Returning to (4.9), it is clear that in the left-half plane Re $\lambda < 0$ and that hence as $|\lambda| \to \infty$, $e^{\varepsilon\lambda}/(\lambda + a) \to 0$ more rapidly than as $1/\lambda$. The integral along I, therefore, vanishes and $\mathscr{L}[f(t; T)]$ is given simply by the summation term of (4.14), as in the case of rational $F(\lambda)$ with $\nu > 1$.

This technique of shifting the discontinuity is commonly employed in connection with engineering problems and it will also be adopted here. Unless otherwise stated, if $\{f(t)\}$ has a discontinuity at a sampling instant $t = kT$, it will be assumed that the discontinuity occurs *just prior* to the sampling instant, that is, at $t = kT - \varepsilon$. On the basis of this assumption, if (4.14) is used to obtain $\mathscr{L}[f(t; T)]$, the term $-f(0_+)/2$ will always be omitted.

We may now let $z = e^{sT}$ in (4.14) and expand each of the terms in the summation into a power series in z^{-1}. After combining coefficients of like powers, we may write

$$\mathscr{L}[f(t; T)] = \sum_{k=0}^{\infty} f(kT)z^{-k} \Big|_{z=e^{sT}} \tag{4.15}$$

where it is assumed that any discontinuity near $t = 0$ occurs at $t = 0_-$. This result, except for the substitution $z = e^{sT}$, is identical with that of (3.98), and it suggests the notion of the "z transform of a continuous-time function." The latter is merely a compact way of referring to the z transform of the time sequence resulting from sampling $\{f(t)\}$ at $t = kT$.

[4] See G. Doetsch, *Theorie und Anwendung der Laplace Transformation*, Dover Publications, New York, 1943, pp. 104–109; and R. V. Churchill, *Operational Mathematics* (2nd ed.), McGraw-Hill Book Co., New York, 1958, pp. 181–183.

Thus

$$\mathcal{Z}[f(t)] = F_T(z) = \sum \text{residues of } \frac{F(\lambda)}{1 - e^{T\lambda}z^{-1}} \quad \text{at poles of } F(\lambda) \tag{4.16}$$

Note from (4.15) and (4.16) that we can obtain the z transform of the time sequence obtained by sampling the continuous function $\{f(t)\}$ in either of two ways. We may find $f(kT)$ from $\{f(t)\}$ and then obtain the z transform according to (3.23), or we may first obtain $F(\lambda)$ and then apply (4.16). The result will, of course, be the same.

Example

Find the z transform of $f(t) = 1 + a^t$, $t \geq 0$ in two different ways.

(*a*)
$$f(kT) = 1 + a^{kT}$$

$$\begin{aligned} F_T(z) &= \sum_{k=0}^{\infty} f(kT)z^{-k} \\ &= (1 + 1) + (1 + a^T)z^{-1} + (1 + a^{2T})z^{-2} + \cdots \\ &= \frac{1}{1 - z^{-1}} + \frac{1}{1 - a^Tz^{-1}} \qquad \text{for } |z| > \max(1, a^T) \end{aligned}$$

(*b*)
$$\begin{aligned} F(\lambda) &= \frac{1}{\lambda} + \frac{1}{\lambda - \log a} \\ &= \frac{2\lambda - \log a}{\lambda(\lambda - \log a)} \end{aligned}$$

Using (4.16), we have

$$F_T(z) = \sum \text{res} \frac{2\lambda - \log a}{\lambda(\lambda - \log a)(1 - e^{T\lambda}z^{-1})}$$

$$\text{poles:} \quad \lambda = 0 \quad \text{and} \quad \lambda = \log a$$

$$= \frac{1}{1 - z^{-1}} + \frac{1}{1 - a^Tz^{-1}} \qquad \text{for } |z| > \max(1, a^T)$$

4.3 ALTERNATE FORM OF $\mathscr{L}[f(t; T)]$

We shall now consider the evaluation of $\mathscr{L}[f(t; T)]$ over the right-half plane in Fig. 4-1, as given by (4.10). The integral over the closed contour II + $\uparrow$ encircles (in a clockwise sense) the poles of $1/(1 - e^{T(\lambda - s)})$. There are an infinite number of isolated poles, located respectively at

$$\lambda_l = s + \frac{2\pi jl}{T}, \qquad l = 0, \pm 1, \pm 2, \ldots, \pm\infty \tag{4.17}$$

that is, the poles are successively spaced by an amount $2\pi/T$ along the line $\lambda = \text{Re } s$ in the λ plane. If $F(\lambda) \to 0$ as $|\lambda| \to \infty$ in the right-half plane, the residues at these poles converge uniformly to zero as $l \to \pm\infty$. Hence by choosing a sufficiently large radius for the semicircular contour II, the contributions from the poles lying outside the contour can be made arbitrarily small. Accordingly:

$$\frac{1}{2\pi j}\oint_{\text{II}+\uparrow} \frac{F(\lambda)\,d\lambda}{1-e^{T(\lambda-s)}} = -\sum_{l=-\infty}^{\infty} \frac{F(\lambda_l)}{-Te^{T(\lambda_l-s)}} \tag{4.18}$$

where the negative sign in front of the summation is due to the clockwise encirclement of the poles. Upon substitution of (4.17),

$$\frac{1}{2\pi j}\oint_{\text{II}+\uparrow} \frac{F(\lambda)}{1-e^{T(\lambda-s)}}\,d\lambda = \frac{1}{T}\sum_{l=-\infty}^{\infty} F\left(s+\frac{2\pi jl}{T}\right) \tag{4.19}$$

By following analogous reasoning to that used for the evaluation of the integral in (4.12), it is seen that the integral over the semicircular path II vanishes; that is,

$$\frac{1}{2\pi j}\int_{\text{II}} \frac{F(\lambda)}{1-e^{T(\lambda-s)}}\,d\lambda = 0 \tag{4.20}$$

Hence from (4.10) and (4.19),

$$\boxed{\mathscr{L}[f(t;T)] = \frac{1}{T}\sum_{l=-\infty}^{\infty} F\left(s+\frac{2\pi jl}{T}\right)} \tag{4.21}$$

Since the results obtainable from (4.9) clearly must be identical with those from (4.10), we may combine (4.14) and (4.21) and write

$$\boxed{\begin{array}{l}\sum \text{residues of } \dfrac{F(\lambda)}{1-e^{T(\lambda-s)}} - \dfrac{f(0_+)}{2} = \dfrac{1}{T}\displaystyle\sum_{l=-\infty}^{\infty} F\left(s+\frac{2\pi jl}{T}\right) \\ \text{at poles of } F(\lambda)\end{array}} \tag{4.22}$$

In practice the Laplace transform of $\{f(kT)\}$ is more useful when expressed in the form of (4.14) than when given in the form of (4.21). However, the latter does provide some insight into the sampling operation, as will be seen in the next section.

From (4.15) and (4.22),

$$\boxed{\sum_{k=0}^{\infty} f(kT)z^{-k}\bigg|_{z=e^{sT}} = \frac{1}{T}\sum_{l=-\infty}^{\infty} F\left(s+\frac{2\pi jl}{T}\right) + \frac{f(0_+)}{2}} \tag{4.23}$$

Equation (4.23) expresses the equality[5] of the two forms of $\mathscr{L}[f(kT)]$ *for the case where it is assumed that any discontinuity of* $\{f(t)\}$ *at* $t = 0$ *occurs actually at* $t = 0_-$.

4.4 THE SAMPLING THEOREM

An important theorem pertaining to the operation of sampling a continuous-time function is the so-called *sampling theorem.* This theorem states that

"*a function* $\{f(t)\}$ *which has a Fourier spectrum* $F(j\omega)$ *such that* $F(j\omega) = 0$ *for* $|\omega| > \Omega/2$ *is uniquely described by a knowledge of its values at uniformly spaced instants, T units apart* ($T = 2\pi/\Omega$)."

The theorem may be derived as follows.[6] Consider a function $\{f(t)\}$ which possesses a Fourier transform $F(j\omega)$, where

$$F(j\omega) = 0 \quad \text{for } |\omega| > \Omega/2$$

We may stipulate a periodic function $F_1(j\omega)$ which is periodic with period Ω and which is equal to $F(j\omega)$ in the interval $-\Omega/2 < \omega < \Omega/2$. If $F_1(j\omega)$ is expanded in a Fourier series

$$F_1(j\omega) = \sum_{k=-\infty}^{\infty} a_k e^{j\omega kT} \tag{4.24}$$

the series is found to have the coefficients,

$$a_k = \frac{1}{\Omega} \int_{-\Omega/2}^{\Omega/2} F(j\omega) e^{-j\omega kT} \, d\omega \tag{4.25}$$

where

$$T = \frac{2\pi}{\Omega} \tag{4.26}$$

Independently of the foregoing, we may write for the Fourier transform of $F(j\omega)$

$$f(t) = \frac{1}{2\pi} \int_{-\Omega/2}^{\Omega/2} F(j\omega) e^{j\omega t} \, d\omega \tag{4.27}$$

Letting $t = kT$, we have

$$f(kT) = \frac{1}{2\pi} \int_{-\Omega/2}^{\Omega/2} F(j\omega) e^{j\omega kT} \, d\omega \tag{4.28}$$

[5] G. V. Lago, "Addition to Sampled-Data Theory," *Proc. Nat. Electron. Conf.*, vol. 10, pp. 758–766, 1954.

[6] B. M. Oliver, J. R. Pierce, and C. E. Shannon, "The Philosophy of PCM," *Proc. IRE*, vol. 36, no. 11, pp. 1324–1331, November 1948.

Comparing (4.25) and (4.28), we obtain

$$a_k = Tf(-kT) \tag{4.29}$$

Hence if we are given the sample values $f(kT)$ for all integers $-\infty < k < \infty$ and if it is known that $\{f(t)\}$ is bandlimited, that is, $F(j\omega) = 0$ for $|\omega| > \Omega/2 = \pi/T$, then we may uniquely determine the coefficients a_k. Restricting consideration to the interval $-\Omega/2 < \omega < \Omega/2$, we may write

$$\begin{aligned} F(j\omega) &= \sum_{k=-\infty}^{\infty} a_k e^{j\omega kT} \quad &\text{for } |\omega| < \Omega/2 \\ &= 0 &\text{for } |\omega| > \Omega/2 \end{aligned} \tag{4.30}$$

Inverting $F(j\omega)$ yields

$$f(t) = \frac{1}{2\pi} \int_{-\Omega/2}^{\Omega/2} \sum_{k=-\infty}^{\infty} a_k e^{j\omega kT} e^{j\omega t}\, d\omega \tag{4.31}$$

$$= \frac{1}{T} \sum_{k=-\infty}^{\infty} a_k \frac{\sin \Omega(t + kT)/2}{\Omega(t + kT)/2} \tag{4.32}$$

Using (4.29) and replacing k by $-k$, we have

$$\boxed{f(t) = \sum_{k=-\infty}^{\infty} f(kT) \frac{\sin \Omega(t - kT)/2}{\Omega(t - kT)/2}} \tag{4.33}$$

A comparison between (4.33) and (2.6) shows that the former may be interpreted as the convolution summation of the modulated delta-function series $\{f(t; T)\}$ with the filter weighting function

$$g(t) = \frac{\sin \Omega t/2}{\Omega t/2} \tag{4.34}$$

A sketch of this weighting function is given in Fig. 4-2. The function is seen to equal unity at $t = 0$ and to vanish at all instants $t = kT$, $k \neq 0$.

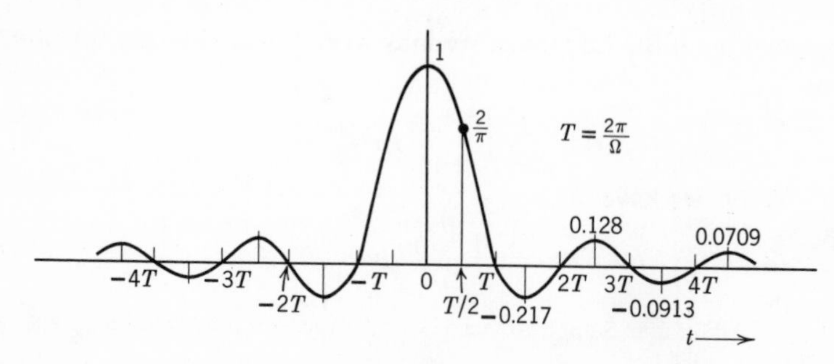

Figure 4-2 Sketch of $\frac{\sin \Omega t/2}{\Omega t/2}$ weighting function.

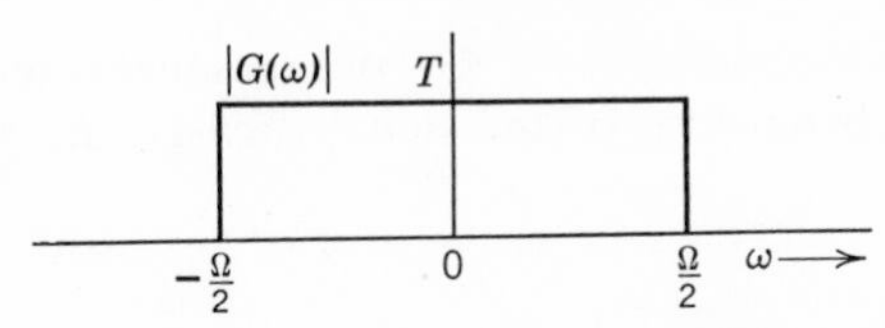

Figure 4-3 Fourier spectrum of $\frac{\sin \Omega t/2}{\Omega t/2}$ filter.

The corresponding Fourier spectrum is shown in Fig. 4-3. The filter is seen to be an ideal bandpass filter, having a constant output-to-input amplitude ratio of T in the interval $-\Omega/2 < \omega < \Omega/2$ and zero elsewhere. Although a filter having this weighting function is not physically realizable (the weighting function is anticipatory), it can be approximated by preceding it with a sufficiently long, pure time delay. Thus if $F(j\omega)$ is bandlimited and if Ω is chosen such that $F(j\omega) = 0$ for $|\omega| \geq \Omega/2$, it is possible (at least in an ideal sense) to recover $\{f(t)\}$ from the samples $f(kT)$, $k = 0, \pm 1, \pm 2, \ldots \to \pm\infty$.

The relation between the sampling theorem and the ideal bandpass filter of Fig. 4-3 can be made even more direct by referring to (4.21). It is seen that the Laplace transform of $\{f(t; T)\}$ is represented by a summation of shifted transforms $F(s + jk\Omega)$, all multiplied by $1/T$. If $F(s)$ is bandlimited such that $F(j\omega) = 0$ for $|\omega| > \Omega/2$, the various terms in the summation will be distinct and nonoverlapping in the complex frequency domain. Hence passing $\mathscr{L}[f(t; T)]$ through the ideal bandpass filter of Fig. 4-3 will cause the term for $l = 0$ to be selected and multiplied by T. All other terms (corresponding to $l \neq 0$) are rejected and the filter output will be precisely $F(s) = \mathscr{L}[f(t)]$. Conversely, if $\{f(t)\}$ does not represent a bandlimited signal, $\mathscr{L}[f(t; T)]$ as given by (4.21) will have overlapping components and it will *not* be possible to recover $\{f(t)\}$.[7]

The sampling theorem defines a minimum value for the sampling frequency Ω, given a particular bandlimited $\{f(t)\}$, below which recovery of $\{f(t)\}$ is impossible. From a practical point of view the sampling frequency must, of course, be many times greater than Ω; in practice, we encounter neither bandlimited signals nor ideal filters. Nor in fact would we have available an infinite set of samples of the function.

When determining the frequency with which a given function is to be sampled, it is important to realize that the bandwidth of the *total signal* must be considered. Thus if a channel contains a signal $\{f(t)\}$ consisting of both a message $\{f_1(t)\}$ which is bandlimited to $-\Omega/2 < \omega < \Omega/2$, and noise $\{f_2(t)\}$, which lies outside of this band (so that the total signal bandwidth is greater than $-\Omega/2 < \omega < \Omega/2$), we can recover *neither* $\{f_1(t)\}$

[7] Not even with a physically nonrealizable filter!

nor $\{f(t)\}$ from the samples $f(kT)$. For this reason it is generally preferable to speak of the bandwidth of the signal *channel* rather than the signal itself.

4.5 EXTENSIONS OF THE SAMPLING THEOREM

It is possible to make certain extensions and generalizations of the sampling theorem. Examination of (4.33) shows that $\{f(t)\}$ can be visualized as being represented by the infinite sum of pulses of form

$$\frac{\sin \Omega(t - kT)/2}{\Omega(t - kT)/2}$$

each multiplied by an appropriate coefficient $f(kT)$. The shape of the pulses is as shown in Fig. 4-2. Observe that the amplitude of the pulse decreases to $2/\pi$ of its peak value at $T/2$ seconds from the point about which it is centered, and decreases more rapidly thereafter. Considering energy content, evaluation of the integral

$$\int_{-T/2}^{T/2} \left(\frac{\sin t}{t}\right)^2 dt$$

shows that nearly 80% of the total pulse energy[8] is contained within $-T/2 \leq t \leq T/2$. It follows that if we desire to describe a bandlimited function $\{f(t)\}$ over a finite time interval of length τ, that is, over τ/T such pulse widths, we require τ/T uniformly spaced samples of $\{f(t)\}$. If the bandwidth of the function $\{f(t)\}$ is given as W cycles per second, the sampling interval T is given by

$$T = \frac{2\pi}{\Omega} = \frac{2\pi}{4\pi W} = \frac{1}{2W} \tag{4.35}$$

Hence

$$\frac{\tau}{T} = 2W\tau \tag{4.36}$$

The foregoing leads to a more general statement of the sampling theorem, according to which *a function* $\{f(t)\}$ *which is bandlimited to the interval* $-W < f < W$ *(in cycles per second) and is time-limited to an interval of length* τ *seconds is completely specified by knowledge of* $2W\tau$ *of its values.*

Since a truly bandlimited signal cannot also be time-limited (and vice versa), the foregoing is clearly an approximation. The approximation

[8] The total energy of the pulse is equal to π.

improves if $\{f(t)\}$ tapers toward zero as it approaches the ends of the time interval τ and if the interval is made progressively larger.[9]

We shall consider here only *sampling expansions* of $\{f(t)\}$, that is, expansions that are of the general form (4.33) and which express $f(t)$ for any t in terms of a countable set of samples of $\{f(t)\}$. As will be shown in the next three sections, these samples need not be either uniformly spaced, or, in fact, values of $\{f(t)\}$ directly. They must merely represent $2W\tau$ *independent pieces of information about* f.

The sampling expansions are all based on use of the so-called *cardinal function* $(\sin t)/t$. As already stated, this function tends to concentrate the major fraction of its energy in the interval $|t| \leq T/2$. Hence if a bandlimited $\{f(t)\}$ is represented by $2W\tau$ such functions over an interval of length τ, only a small fraction of the energy of these functions will lie outside this interval. The energy lying outside the specified time interval can be minimized if we use the so-called *prolate spheroidal wave functions*[10] instead of the cardinal function. These spheroidal wave functions are orthonormal, bandlimited functions into which $\{f(t)\}$ can be expanded in a manner analogous to a Fourier series expansion. It can be shown[11] that if a given bandlimited $\{f(t)\}$ that is essentially time-limited to an interval of length τ is expanded in terms of prolate spheroidal wave functions, $\{f(t)\}$ can be represented to an arbitrary precision by means of $2W\tau$ independent signals. This result is, however, not valid for expansions using the cardinal function. Nevertheless it lends some justification to the loose statement that $2W\tau$ independent signals suffice to describe a signal bandlimited to $(-W, W)$ over a time interval of length τ.

4.6 PERIODIC NONUNIFORM SAMPLING

Consider a bandlimited function $\{f(t)\}$ whose Fourier transform vanishes outside the interval $-(m+1)\Omega/2 < \omega < (m+1)\Omega/2$. The values of $\{f(t)\}$ are given at $t = t_{ki} = kT + t_i$, where k ranges over all integers, $i = 0, 1, \ldots, m$, and $(\max t_i) < T$. The spacing between samples of $\{f(t)\}$ is thus nonuniform, but with a pattern which is periodic with period T.[12] An illustration of this type of nonuniform sampling is shown in Fig. 4-4, where $m = 3$.

[9] C. E. Shannon, "Communications in the Presence of Noise," *Proc. IRE*, vol. 37, no. 1, pp. 10–21, January 1949.

[10] H. J. Landau, and H. O. Pollak, "Prolate Spheroidal Wave Functions, Fourier Analysis and Uncertainly—III: The Dimensions of the Space of Essentially Time— and Bandlimited Signals," *Bell System Tech. J.*, vol. 41, pp. 1295–1336, July 1962.

[11] H. J. Landau, and H. O. Pollak, *op. cit.*, pp. 1295–1303.

[12] J. L. Yen, "On Nonuniform Sampling of Bandwidth-Limited Signals," *IRE Trans. Circuit Theory*, vol. CT-3, pp. 251–257, December 1956.

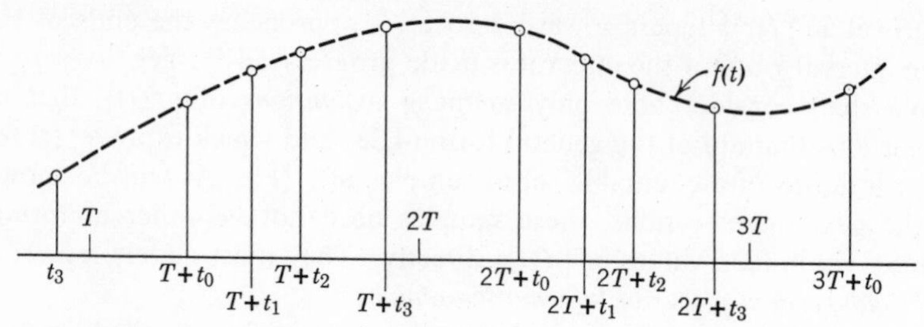

Figure 4-4 Illustration of periodic nonuniform sampling ($m = 3$).

From the given samples $f(t_{ki})$ we can form the modulated delta-function series

$$g_i(t) = \sum_{k=-\infty}^{\infty} f(t)\delta(t - t_{ki}), \qquad i = 0, 1, \ldots, m \tag{4.37}$$

We denote the Fourier transforms of the functions $\{g_i(t)\}$ by $G_i(j\omega)$ and proceed as indicated in Fig. 4-5. Let

$$\mathscr{F}[f(t)] = F(j\omega)$$

Then

$$\mathscr{F}[f(t + t_i)] = e^{j\omega t_i}F(j\omega) \tag{4.38}$$

Since $t_{ki} = kT + t_i$,

$$g_i(t + t_i) = \sum_{k=-\infty}^{\infty} f(t + t_i)\delta(t - kT) \tag{4.39}$$

Using (3.103), (3.104), and (4.21), we obtain from (4.38)

$$\mathscr{F}[g_i(t + t_i)] = \frac{1}{T}\sum_{l=-\infty}^{\infty} e^{j(\omega + l\Omega)t_i}F(j\omega + jl\Omega). \tag{4.40}$$

Hence

$$\begin{aligned} G_i(j\omega) &= e^{-j\omega t_i}\mathscr{F}[g_i(t + t_i)] \\ &= \frac{1}{T}\sum_{l=-\infty}^{\infty} e^{jl\Omega t_i}F(j\omega + jl\Omega), \qquad i = 0, 1, \ldots, m \end{aligned} \tag{4.41}$$

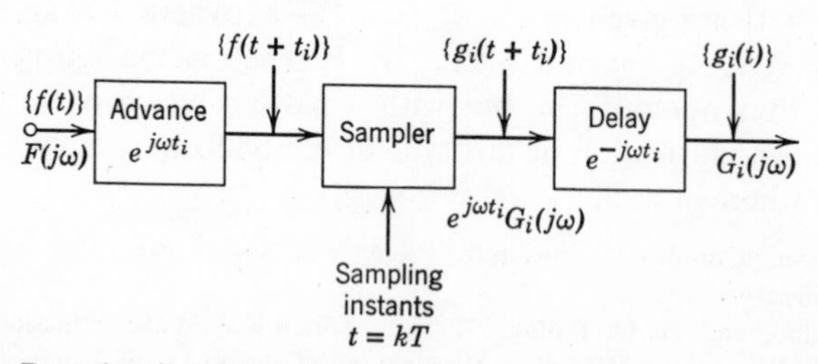

Figure 4-5 Procedure for determining the Fourier transform of the modulated delta-function series $g_i(t) = \sum_{k=-\infty}^{\infty} f(t)\delta(t - t_{ki})$.

We now divide the ω interval for which $F(j\omega)$ may be nonzero, that is, $-(m+1)\Omega/2 < \omega < (m+1)\Omega/2$, into $2(m+1)$ equal parts, each of length $\Omega/2$. Let

$$F_q(j\omega) = F(j\omega) \quad \text{for } \frac{q\Omega}{2} \leq \omega < \frac{(q+1)\Omega}{2}$$
$$= 0 \qquad \text{elsewhere} \tag{4.42}$$

where q is an integer, $-(m+1) \leq q \leq m$.

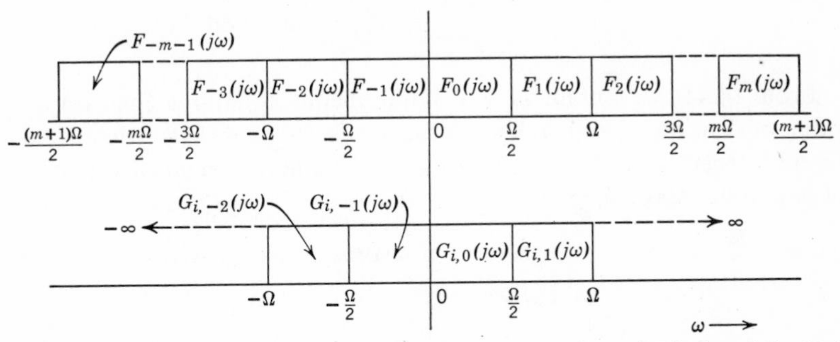

Figure 4-6 Diagram of the ω intervals for the nonzero values of $F_q(j\omega)$ and $G_{i,p}(j\omega)$.

Also let

$$G_{i,p}(j\omega) = G_i(j\omega) \quad \text{for } p\Omega/2 \leq \omega < (p+1)\Omega/2$$
$$= 0 \qquad \text{elsewhere} \tag{4.43}$$

where p may be any integer, $-\infty < p < \infty$.

Since $F(j\omega) = 0$ for $|\omega| > (m+1)\Omega/2$, each $G_{i,p}(j\omega)$ can be expressed as the sum of a *finite* number of terms of $F_q(j\omega)$. This may be seen most clearly by referring to Fig. 4-6. Thus from (4.41) we may write for $G_{i,0}(j\omega)$

$$G_{i,0}(j\omega) = \frac{1}{T}\sum_{l=\alpha_0}^{\alpha_0+m} e^{jl\Omega t_i} F_{2l}(j\omega + jl\Omega) \tag{4.44}$$

where

$$\alpha_0 = \frac{-m}{2} \qquad \text{if } m \text{ is even}$$
$$= -\frac{1+m}{2} \qquad \text{if } m \text{ is odd} \tag{4.45}$$

For example, for $m = 2$,

$$G_{i,0}(j\omega)|_{m=2} = \frac{1}{T}\left[e^{-j\Omega t_i}F_{-2}(j\omega - j\Omega) + F_0(j\omega) + e^{j\Omega t_i}F_2(j\omega + j\Omega)\right]$$

and for $m = 3$,

$$G_{i,0}(j\omega)|_{m=3} = \frac{1}{T}\left[e^{-j2\Omega t_i}F_{-4}(j\omega - 2j\Omega) + e^{-j\Omega t_i}F_{-2}(j\omega - j\Omega) + F_0(j\omega) + e^{j\Omega t_i}F_2(j\omega + j\Omega)\right]$$

Equations such as (4.44) may, of course, be written for *any* value of p. In general,

$$G_{i,p}(j\omega) = \frac{1}{T}\sum_{l=\alpha_p}^{\alpha_p+m} e^{jk\Omega t_i} F_{p+2l}(j\omega + jl\Omega) \tag{4.46}$$

where

$$\begin{aligned} \alpha_p &= -\frac{p+m}{2} \qquad \text{if } p+m \text{ is even} \\ &= -\frac{p+m+1}{2} \qquad \text{if } p+m \text{ is odd} \end{aligned} \tag{4.47}$$

From (4.44) we obtain $m + 1$ simultaneous equations (one for each value of i) in the $m + 1$ unknown functions $F_{2l}(j\omega + jl\Omega)$, where l takes on all integer values from α_0 to $\alpha_0 + m$. The determinant of this system of equations is as follows:

$$\Delta = \left(\frac{1}{T}\right)^{m+1} \begin{vmatrix} e^{j\alpha_0\Omega t_0} & e^{j(\alpha_0+1)\Omega t_0} & \cdots & e^{j(\alpha_0+m)\Omega t_0} \\ e^{j\alpha_0\Omega t_1} & e^{j(\alpha_0+1)\Omega t_1} & \cdots & e^{j(\alpha_0+m)\Omega t_1} \\ \cdot & \cdot & & \cdot \\ \cdot & \cdot & & \cdot \\ \cdot & \cdot & & \cdot \\ e^{j\alpha_0\Omega t_m} & e^{j(\alpha_0+1)\Omega t_m} & \cdots & e^{j(\alpha_0+m)\Omega t_m} \end{vmatrix} \tag{4.48}$$

$$= \frac{\prod_{i=0}^{m} e^{j\alpha_0\Omega t_i}}{T^{m+1}} \begin{vmatrix} 1 & e^{j\Omega t_0} & \cdots & e^{jm\Omega t_0} \\ 1 & e^{j\Omega t_1} & & e^{jm\Omega t_1} \\ \cdot & \cdot & & \cdot \\ \cdot & \cdot & & \cdot \\ \cdot & \cdot & & \cdot \\ 1 & e^{j\Omega t_m} & \cdots & e^{jm\Omega t_m} \end{vmatrix} \tag{4.49}$$

In this latter form the determinant is recognized as the so-called Vandermonde determinant, for which the expansion is available.[13] We are thus able to write

$$\Delta = \prod_{i=0}^{m} e^{j\alpha_0\Omega t_i} \cdot \prod_{r>s\geq 0}^{m} [e^{j\Omega t_r} - e^{j\Omega t_s}] T^{-(m+1)} \tag{4.50}$$

All the sampling instants t_i are distinct and (max t_i) $< T$ by hypothesis. Hence none of the factors of (4.50) vanish and Δ is nonsingular. Therefore, from Cramer's rule, the solution of the $m + 1$ equations implied by (4.44)

[13] See, for example, G. Birkhoff and S. MacLane, *A Survey of Modern Algebra*, Macmillan Co., New York, 1950, p. 288; or R. R. Stoll, *Linear Algebra and Matrix Theory*, McGraw-Hill Book Co., New York, 1952, p. 102.

is unique. The $m+1$ functions $F_{2l}(j\omega + jl\Omega)$ are sufficient to determine all the $F_q(j\omega)$ since, by symmetry,

$$F_q(j\omega) = F^*_{-q-1}(-j\omega) \tag{4.51}$$

where the asterisk indicates the complex conjugate. A completely equivalent result is obtainable from the solution of (4.46) for *any* particular value of p.

We thus conclude that a function $\{f(t)\}$ bandlimited to the interval $-(m+1)\Omega/2 < \omega < (m+1)\Omega/2$ can be uniquely recovered from a set of samples $f(t_{ki})$ obtained at instants $t = t_{ki} = kT + t_i$, spaced according to a periodically recurring, nonuniform pattern.

Having established both the possibility as well as the uniqueness of the recovery of $\{f(t)\}$ from $f(t_{ki})$, we may now seek out the interpolation function which will actually accomplish this. What is desired is a formula analogous to (4.33).

We may write

$$f(t) = \sum_{k=-\infty}^{\infty} \sum_{i=0}^{m} f(t_{ki})\varphi_{ki}(t) \tag{4.52}$$

where φ_{ki} is the desired interpolation function. The requirements on φ_{ki} are as follows:

For integer values of r and s,

$$\begin{aligned} \varphi_{ki}(t_{rs}) &= 1 \quad \text{if } r = k \text{ and } s = i \\ &= 0 \quad \text{if } r \neq k \text{ or } s \neq i \end{aligned}$$

and the Fourier transform of φ_{ki} must vanish outside the interval

$$-(m+1)\Omega/2 < \omega < (m+1)\Omega/2$$

Consider the form

$$\varphi_{ki}(t) = \frac{\displaystyle\prod_{v=0}^{m} \sin \frac{\Omega(t - kT - t_v)}{2}}{\dfrac{K\Omega(t - kT - t_i)}{2}} \tag{4.53}$$

where K is a scalar. Clearly, the Fourier spectrum of this expression is governed by the $m+1$ sine function terms of the numerator. Since each term is of radian frequency $\Omega/2$, their product is of bandwidth $(m+1)\Omega/2$. Hence the expression satisfies the requirement that its Fourier transform vanish outside the interval $-(m+1)\Omega/2 < \omega < (m+1)\Omega/2$. Further, the right-hand side of the equation vanishes for all $t = kT + t_v$ except

$t = kT + t_i$, $\nu = 0, 1, \ldots, m$. To assure that the expression takes on the value unity at $t = kT + t_i$, it is only necessary that

$$K = \prod_{\substack{\nu=0 \\ \nu \neq i}}^{m} \sin \frac{\Omega(t_i - t_\nu)}{2}.$$

Thus the desired interpolation formula is as follows:[14]

$$\boxed{f(t) = \sum_{k=-\infty}^{\infty} \sum_{i=0}^{m} f(t_{ki}) \frac{\displaystyle\prod_{\nu=0}^{m} \sin \frac{\Omega(t - kT - t_\nu)}{2}}{\displaystyle\frac{\Omega(t - kT - t_i)}{2} \prod_{\substack{\nu=0 \\ \nu \neq i}}^{m} \sin \frac{\Omega(t_i - t_\nu)}{2}}} \tag{4.54}$$

4.7 UNIFORM SAMPLING OF A FUNCTION AND ITS DERIVATIVES

It is also possible to apply the sampling theorem when uniformly spaced samples of the function as well as of the function's derivatives are given.[15] Consider a function $\{f(t)\}$ which is bandlimited to $-(m + 1)\Omega/2 < \omega < (m + 1)\Omega/2$ and for which both $f(kT)$ as well as the samples of the first m derivatives are given, that is, $f^{(i)}(kT)$ for $i = 0, 1, \ldots, m$, $T = 2\pi/\Omega$.

Let

$$\mathscr{F}[f(t)] = F(j\omega) \tag{4.55}$$

Then

$$\mathscr{F}[f^{(i)}(t)] = (j\omega)^i F(j\omega) \tag{4.56}$$

$$H_i(j\omega) = \mathscr{F}[f^{(i)}(kT)] = \frac{1}{T} \sum_{l=-\infty}^{\infty} (j\omega + jl\Omega)^i F(j\omega + jl\Omega), \qquad i = 0, 1, 2, \ldots, m \tag{4.57}$$

We again divide the ω interval for which $F(j\omega)$ may be nonzero, that is, $-(m + 1)\Omega/2 < \omega < (m + 1)\Omega/2$, into $2(m + 1)$ equal parts, each of length $\Omega/2$, as shown in Fig. 4-6. Let $F_q(j\omega)$ be as defined by (4.42). Also let

$$\begin{aligned} H_{i,p}(j\omega) &= H_i(j\omega) \quad \text{for } p\Omega/2 \leq \omega < (p + 1)\Omega/2 \\ &= 0 \qquad \text{elsewhere} \end{aligned} \tag{4.58}$$

[14] It has been shown by Helms that it is possible to obtain interpolation formulas of this type in a more direct manner: H. D. Helms, "Generalizations of the Sampling Theorem and Error Calculations," Ph.D. dissertation, Princeton University, January 1961.

[15] D. A. Linden and N. M. Abramson, "A Generalization of the Sampling Theorem," *Inform. Control*, vol. 3, no. 1, pp. 26–31, March 1960.

where p may be any integer, $-\infty < p < \infty$. In Fig. 4-6, each $G_{i,p}(j\omega)$ should now be replaced by the corresponding $H_{i,p}(j\omega)$.

Proceeding in accordance with the method used in the determination of (4.44) and (4.46), we write

$$H_{i,p}(j\omega) = \frac{1}{T}\sum_{l=\alpha_p}^{\alpha_p+m}(j\omega + jl\Omega)^i F_{p+2l}(j\omega + jl\Omega), \qquad i = 0, 1, \ldots, m \tag{4.59}$$

where α_p is as given by (4.47); that is,

$$\alpha_p = -\frac{p+m}{2} \qquad \text{if } p + m \text{ is even}$$

$$= -\frac{p+m+1}{2} \qquad \text{if } p + m \text{ is odd}$$

The $m + 1$ unknown components $F_q(j\omega)$ of $F(j\omega)$ can be determined by solving a set of $m + 1$ simultaneous equations obtainable from (4.59) by selecting a particular value of p. Let $p = 0$. Then we obtain $m + 1$ equations (one for each value of i) in the $m + 1$ unknown functions $F_{2l}(j\omega + jl\Omega)$, where l takes on all integer values from α_0 to $\alpha_0 + m$. The determinant of this set of equations is as follows:

$$\Delta = \left(\frac{1}{T}\right)^{m+1}\begin{vmatrix} 1 & 1 & \cdots & 1 \\ j\omega + j\alpha_0\Omega & j\omega + j(\alpha_0 + 1)\Omega & \cdots & j\omega + j(\alpha_0 + m)\Omega \\ [j\omega + j\alpha_0\Omega]^2 & [j\omega + j(\alpha_0 + 1)\Omega]^2 & \cdots & [j\omega + j(\alpha_0 + m)\Omega]^2 \\ \vdots & \vdots & & \vdots \\ [j\omega + j\alpha_0\Omega]^m & [j\omega + j(\alpha_0 + 1)\Omega]^m & \cdots & [j\omega + j(\alpha_0 + m)\Omega]^m \end{vmatrix} \tag{4.60}$$

This determinant is of the same form as (4.49) and its expansion yields

$$\Delta = \left(\frac{1}{T}\right)^{m+1}\prod_{r>s\geq 0}^{m}\{[j\omega + j(\alpha_0 + r)\Omega] - [j\omega + j(\alpha_0 + s)\Omega]\} \tag{4.61}$$

$$= T^{-(m+1)}\prod_{r>s\geq 0}^{m}[j\Omega(r - s)] \tag{4.62}$$

Since $r > s$, none of the factors of the determinant vanishes and hence the determinant is nonsingular. The system of simultaneous equations represented by (4.59) for a particular value of p thus uniquely yields all the $F_i(j\omega)$ required to define $F(j\omega)$.

To find the interpolation formula that will permit the reconstruction of $\{f(t)\}$ from the samples $f^{(i)}(kT)$, $i = 0, 1, \ldots, m$ and $k = 0, \pm 1, \pm 2, \ldots, \pm\infty$, it is merely necessary to find a function $\{f_a(t)\}$ such that

$$\text{(1)} \qquad f_a^{(i)}(kT) = f^{(i)}(kT) \quad \text{for all } i \text{ and } k \tag{4.63}$$

and

$$\text{(2)} \qquad \mathscr{F}[f_a(t)] = 0 \qquad \text{for } |\omega| > (m+1)\Omega/2 \tag{4.64}$$

Since, in accordance with the foregoing, (1) and (2) define a unique function of time, $f_a(t) \equiv f(t)$ for all t.

Consider the following formula:

$$\boxed{f(t) = \sum_{k=-\infty}^{\infty} \sum_{i=0}^{m} \frac{(t-kT)^i}{i!} f^{(i)}(kT) \left[\frac{\sin \Omega(t-kT)/2}{\Omega(t-kT)/2}\right]^{m+1}} \tag{4.65}$$

Successive differentiation shows that (4.65) gives the correct values of $\{f^{(i)}(t)\}$ at $t = kT$ for all $i = 0, 1, \ldots, m$ and all integers k, $-\infty < k < \infty$. However, before (4.65) can be accepted as the correct (and unique) interpolation formula for $\{f(t)\}$, it is also necessary to show that the expression on the right in (4.65) has a Fourier transform that vanishes for $|\omega| > (m+1)\Omega/2$.

By analogy with (2.6), (4.65) can be regarded as the sum of $m+1$ convolutions of the delta-function series $\{f^{(i)}(kT)\}$ with the filter weighting function

$$t^i \left[\frac{\sin \Omega t/2}{\Omega t/2}\right]^{m+1}$$

Since the factor t^i does not affect the bandwidth of this filter, we need concern ourselves only with the bandwidth of

$$w_m(t) = \left[\frac{\sin \Omega t/2}{\Omega t/2}\right]^{m+1} \tag{4.66}$$

The Fourier transform of $\{w_m(t)\}$ is given by the m-fold convolution[16] of the Fourier transform $W(j\omega)$ of the ideal bandpass filter described in Fig. 4-3. Since the Fourier spectrum of the ideal bandpass filter is zero outside the interval $-\Omega/2 < \omega < \Omega/2$, the Fourier spectrum of $\{w_m(t)\}$ will be zero outside the interval $-(m+1)\Omega/2 < \omega < (m+1)\Omega/2$. For example, for $m = 1$,

$$\mathscr{F}[w_1(t)] = W_1(j\omega) = \frac{1}{2\pi}\int_{-\infty}^{\infty} W(j\lambda)W(j\omega - j\lambda)\, d\lambda \tag{4.67}$$

where

$$\begin{aligned} |W(j\lambda)| &= T \qquad \text{for } -\Omega/2 \le \omega \le \Omega/2 \\ &= 0 \qquad \text{elsewhere} \end{aligned} \tag{4.68}$$

[16] See (4.5) for the analogous Laplace transform convolution.

It follows that

$$|W_1(j\omega)| = T^2\left(1 - \frac{|\omega|}{\Omega}\right) \qquad \text{for } -\Omega \leq \omega \leq \Omega$$
$$= 0 \qquad \text{elsewhere} \qquad (4.69)$$

A sketch of $|W_1(j\omega)|$ vs. ω is shown in Fig. 4-7.

We conclude that the expression on the right side of (4.65) is band-limited to $-(m+1)\Omega/2 < \omega < (m+1)\Omega/2$. Hence (4.65) is the desired

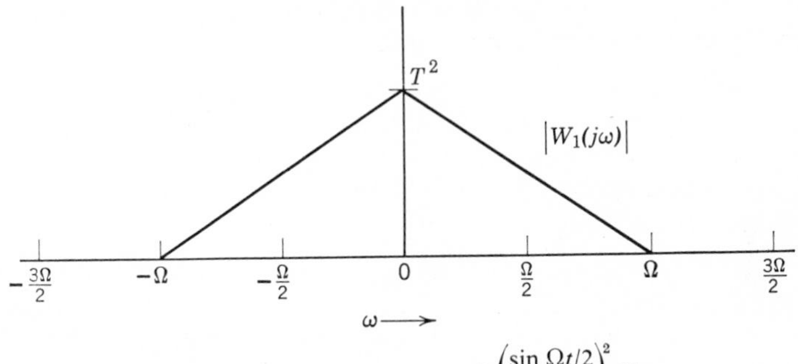

Figure 4-7 Fourier spectrum of $\left(\frac{\sin \Omega t/2}{\Omega t/2}\right)^2$ filter.

formula for the unique reconstruction of $\{f(t)\}$ from the samples $f^{(i)}(kT)$, where $i = 0, 1, \ldots, m$ and where k ranges over all integers from $-\infty$ to ∞.

4.8 NONUNIFORM SAMPLING

The analysis of Section 4.6 was concerned with sampling at instants which, though spaced nonuniformly, repeated with a pattern that was periodic. In this section, attention will be given to the case of non-uniform sampling in a more general sense.

In Fig. 4-8, a set of samples $f(kT)$ of a function $\{f(t)\}$ are shown. Let the Fourier transform $F(j\omega)$ of $\{f(t)\}$ be such that

$$F(j\omega) = 0 \quad \text{for } |\omega| > \Omega/2$$

where $\Omega = 2\pi/T$. Clearly, $\{f(t)\}$ for all t will be completely specified by $f(kT)$ for all integers k. Assume now the situation, illustrated in Fig. 4-8, where for some particular k, say $k = i + 2, f(kT)$ is not known. Instead a value $f(t_{i+2})$ is available, where t_{i+2}/T is not an integer. One asks the question whether the presence of $f(t_{i+2})$ is both necessary and sufficient to make up for the absence of $f(iT + 2T)$, or, more generally, whether

$\{f(t)\}$ can be uniquely reconstructed from a set of samples $f(kT)$ from which a total of ν samples are eliminated and replaced by ν arbitrarily spaced samples $f(t_p)$, t_p/T not an integer ($p = 1, 2, \ldots, \nu$). To answer this question we proceed as follows:[17]

From (4.33),

$$f(t) = \sum_{\substack{k=-\infty \\ k \neq k_p}}^{\infty} f(kT) \frac{\sin \Omega(t - kT)/2}{\Omega(t - kT)/2} + \sum_{p=1}^{\nu} f(k_p T) \frac{\sin \Omega(t - k_p T)/2}{\Omega(t - k_p T)/2} \tag{4.70}$$

where the left summation contains the values of $f(kT)$ which are *known* and the right summation contains those which are *unknown*. Moreover

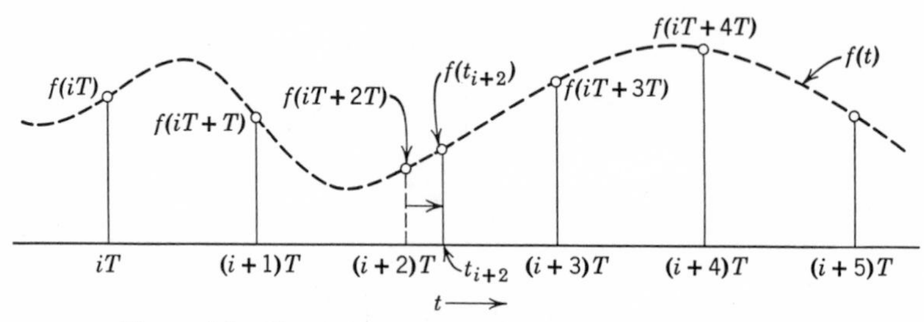

Figure 4-8 Illustration of one type of nonuniform sampling.

in addition to the known values $f(kT)$, $\{f(t)\}$ is also known at ν arbitrarily spaced instants $t = t_p$, $p = 1, 2, \ldots, \nu$. We may thus write ν simultaneous linear equations involving ν unknowns, $f(k_p T)$:

$$\begin{aligned}
\sum_{p=1}^{\nu} f(k_p T) \frac{\sin \Omega(t_1 - k_p T)/2}{\Omega(t_1 - k_p T)/2} &= f(t_1) - \sum_{\substack{k=-\infty \\ k \neq k_p}}^{\infty} f(kT) \frac{\sin \Omega(t_1 - kT)/2}{\Omega(t_1 - kT)/2} \\
\sum_{p=1}^{\nu} f(k_p T) \frac{\sin \Omega(t_2 - k_p T)/2}{\Omega(t_2 - k_p T)/2} &= f(t_2) - \sum_{\substack{k=-\infty \\ k \neq k_p}}^{\infty} f(kT) \frac{\sin \Omega(t_2 - kT)/2}{\Omega(t_2 - kT)/2} \\
&\vdots \\
&\text{etc.}
\end{aligned} \tag{4.71}$$

Now, since k_p is an integer and since $\Omega T = 2\pi$,

$$\begin{aligned}
\sin \frac{\Omega k_p T}{2} &= 0 \\
\cos \frac{\Omega k_p T}{2} &= (-1)^{k_p}
\end{aligned} \tag{4.72}$$

[17] J. L. Yen, *op. cit.*

Hence

$$\frac{\sin \Omega(t_q - k_pT)/2}{\Omega(t_q - k_pT)/2} = \frac{(-1)^{k_p} \sin \Omega t_q/2}{\Omega(t_q - k_pT)/2} \tag{4.73}$$

A typical equation of the set of simultaneous equations (4.71) thus becomes

$$\frac{\sin \Omega t_q/2}{\Omega/2} \sum_{p=1}^{\nu} f(k_pT) \frac{(-1)^{k_p}}{t_q - k_pT} = f(t_q) - \sum_{\substack{k=-\infty \\ k \neq k_p}}^{\infty} f(kT) \frac{\sin \Omega(t_q - kT)/2}{\Omega(t_q - kT)/2} \tag{4.74}$$

Accordingly the determinant of (4.71) can be written as

$$\Delta = \prod_{q=1}^{\nu} \frac{\sin \Omega t_q/2}{(\Omega/2)^{\nu}} \begin{vmatrix} \frac{(-1)^{k_1}}{t_1 - k_1T} & \frac{(-1)^{k_2}}{t_1 - k_2T} & \cdots & \frac{(-1)^{k_\nu}}{t_1 - k_\nu T} \\ \frac{(-1)^{k_1}}{t_2 - k_1T} & \frac{(-1)^{k_2}}{t_2 - k_2T} & \cdots & \frac{(-1)^{k_\nu}}{t_2 - k_\nu T} \\ \vdots & \vdots & & \vdots \\ \frac{(-1)^{k_1}}{t_\nu - k_1T} & \frac{(-1)^{k_2}}{t_\nu - k_2T} & \cdots & \frac{(-1)^{k_\nu}}{t_\nu - k_\nu T} \end{vmatrix} \tag{4.75}$$

The determinant can now be expanded by means of the following formula:[18]

$$\left| \frac{1}{a_r + b_s} \right| = \frac{\prod_{r>s\geq 1}^{\nu} (a_r - a_s)(b_r - b_s)}{\prod_{r,s=1}^{\nu} (a_r + b_s)} \tag{4.76}$$

Thus

$$\Delta = (-1)^{\gamma} \left[\prod_{q=1}^{\nu} \frac{\sin \Omega t_q/2}{(\Omega/2)^{\nu}} \right] \frac{\prod_{q>p\geq 1}^{\nu} (t_q - t_p)(k_pT - k_qT)}{\prod_{q,p=1}^{\nu} (t_q - k_pT)} \tag{4.77}$$

where

$$\gamma = \sum_{p=1}^{\nu} k_p \tag{4.78}$$

Since by hypothesis t_q/T cannot be an integer, $\sin \Omega t_q/2$ will be nonzero for all q. Also for $q \neq p$, $t_p \neq t_q$ and $k_p \neq k_q$. Hence none of the terms in (4.77) can be zero and the determinant is nonsingular. It follows that

[18] E. H. Linfoot and W. M. Shepherd, "On a Set of Linear Equations II," *Quart. J. Math.*, vol. 10, pp. 85–98, 1939.

the solution of the set of simultaneous equations (4.71) will yield a *unique* set of values $f(k_pT)$, $p = 1, 2, \ldots, \nu$.

The derivation initially assumed a uniform distribution from which a finite number of samples were deleted and replaced by a corresponding number of arbitrarily spaced samples. We conclude from this that *a function* $\{f(t)\}$, *bandlimited to* $-\Omega/2 \leq \omega \leq \Omega/2$, *can be uniquely reconstructed from a set of samples which are nonuniformly spaced but satisfy the condition that there be precisely N distinct samples to every interval of length NT, where N is some finite integer.*

The determination of the interpolation function is now a simple task—it is merely necessary to find a function that is bandlimited to $-\Omega/2 < \omega < \Omega/2$ and that takes on the correct values at sampling instants. The foregoing showed that a function which satisfies these conditions will uniquely define $\{f(t)\}$ for all t.

The interpolation formula will be of the form

$$\boxed{f(t) = \sum_{\substack{k=-\infty \\ k \neq k_p}}^{\infty} f(kT)\varphi_a(k, t) + \sum_{p=1}^{\nu} f(t_p)\varphi_b(k_p, t)} \tag{4.79}$$

where $\{\varphi_a(k, t)\}$ is the interpolation function associated with the samples occurring at $t = kT$, and $\{\varphi_b(k_p, t)\}$ is the one associated with the ν samples occurring at $t = t_p$, where t_p/T is not an integer.

Consider first $\{\varphi_a(k, t)\}$. We begin with the term $\dfrac{\sin \Omega(t - kT)/2}{\Omega(t - kT)/2}$. This term has a zero at every $t = lT$, $l \neq k$, but does not vanish at the ν arbitrarily located sampling instants t_p. Hence we must supply ν zeros, one for each t_p. Also it is necessary to remove the zeros corresponding to k_p, that is, the uniformly spaced instants $t = k_pT$ for which samples are *not* available. We thus obtain the form

$$\varphi_a(k, t) = \frac{K_1 \sin \dfrac{\Omega(t - kT)}{2} \displaystyle\prod_{p=1}^{\nu} (t - t_p)}{\dfrac{\Omega(t - kT)}{2} \displaystyle\prod_{p=1}^{\nu} (t - k_pT)} \tag{4.80}$$

The constant K_1 must be selected to give

$$\varphi_a(k, t) = 1 \quad \text{for } t = kT, k \neq k_p$$

Clearly,

$$K_1 = \frac{\displaystyle\prod_{p=1}^{\nu}(kT - k_pT)}{\displaystyle\prod_{p=1}^{\nu}(kT - t_p)} \tag{4.81}$$

and hence

$$\varphi_a(k, t) = \frac{\sin\dfrac{\Omega(t - kT)}{2}\left[\prod_{p=1}^{\nu}(t - t_p)\right]\left[\prod_{p=1}^{\nu}(kT - k_pT)\right]}{\dfrac{\Omega(t - kT)}{2}\left[\prod_{p=1}^{\nu}(t - k_pT)\right]\left[\prod_{p=1}^{\nu}(kT - t_p)\right]}. \tag{4.82}$$

Similarly, for $\{\varphi_b(k_p, t)\}$ we must select a function which will vanish at $t = kT$ for all k except $k = k_p$, and at all $t = t_q$ except for $q = p$. Thus

$$\varphi_b(k_p, t) = \frac{K_2 \sin\dfrac{\Omega t}{2}\prod_{\substack{q=1\\q\neq p}}^{\nu}(t - t_q)}{\prod_{p=1}^{\nu}(t - k_pT)}. \tag{4.83}$$

The constant K_2 must be selected to give

$$\varphi_b(k_p, t) = 1 \quad \text{for } t = t_p$$

We obtain

$$K_2 = \frac{\prod_{p=1}^{\nu}(t_p - k_pT)}{\sin\dfrac{\Omega t_p}{2}\prod_{\substack{q=1\\q\neq p}}^{\nu}(t_p - t_q)}. \tag{4.84}$$

Hence

$$\varphi_b(k_p, t) = \frac{\sin\dfrac{\Omega t}{2}\left[\prod_{\substack{q=1\\q\neq p}}^{\nu}(t - t_q)\right]\left[\prod_{p=1}^{\nu}(t_p - k_pT)\right]}{\sin\dfrac{\Omega t_p}{2}\left[\prod_{\substack{q=1\\q\neq p}}^{\nu}(t_p - t_q)\right]\left[\prod_{p=1}^{\nu}(t - k_pT)\right]} \tag{4.85}$$

The three equations, (4.79), (4.82) and (4.85), thus permit the reconstruction of a bandlimited $\{f(t)\}$ from samples which are nonuniformly spaced but which occur so that in every interval of length NT, where N is some finite integer, there are precisely N samples.

4.9 $\mathcal{Z}$ TRANSFORMS FOR SHIFTED TIME FUNCTIONS

In Section 4.2 it was shown how we may obtain the z transform of the time sequence resulting from sampling a function $\{f(t)\}$ at instants $t = kT$, where $f(t) = 0$ for $t < 0$. It is also of interest to find an expression for the

z transform of a *shifted* continuous-time function, $\{f(t - \tau)\}$, where the z transform of $\{f(t)\}$ sampled at $t = kT$, $k = 0, 1, 2, \ldots$, is given by $F_T(z)$.

If τ is an integer multiple of the sampling period T, we obtain, in accordance with (3.43),

$$\mathcal{Z}[f(t - lT)] = z^{-l}F_T(z) \tag{4.86}$$

If τ is not an integer multiple of T, let

$$\tau = lT - \delta T \tag{4.87}$$

where l is an integer and $0 \leq \delta \leq 1$. We may then write for the z transform

$$\mathcal{Z}[f(t - \tau)] = \sum_{k=0}^{\infty} f(kT + \delta T)z^{-(k+l)} \tag{4.88}$$

The coefficients of the power series expansion in z^{-1} are thus samples of $\{f(t)\}$ taken at $t = (k + \delta)T$ instead of kT. Since $\{f(t)\}$ is defined for the continuous variable t, the values $f(kT + \delta T)$ exist and can be obtained directly from $\{f(t)\}$.

A closed expression for the z transform (4.88) can be obtained most directly by utilizing the integral (4.8). Let

$$\mathcal{L}[f(t)] = F(\lambda) \tag{4.89}$$

Then

$$\mathcal{L}[f(t - \tau)] = e^{-(lT-\delta T)\lambda}F(\lambda) \tag{4.90}$$

Replacing e^{sT} by z in (4.8), we now write

$$\mathcal{Z}[f(t - \tau)] = F_T(z, \tau) = \frac{1}{2\pi j}\int_{c-j\infty}^{c+j\infty} d\lambda\, \frac{e^{-(lT-\delta T)\lambda}F(\lambda)}{1 - e^{\lambda T}z^{-1}} \tag{4.91}$$

Unfortunately, because of the presence of the factor $e^{-lT\lambda}$, this integrand does not converge in the left-half λ plane as $|\lambda|$ becomes infinitely large. The difficulty is avoided by recognizing that the delay τ can be regarded as a delay lT (where l is an integer) less an "advance" δT. Then, utilizing the result of (4.86), we obtain

$$\boxed{F_T(z, \tau) = z^{-l}\frac{1}{2\pi j}\int_{c-j\infty}^{c+j\infty} d\lambda\, \frac{e^{\delta T\lambda}F(\lambda)}{1 - e^{\lambda T}z^{-1}}, \qquad \tau = lT - \delta T} \tag{4.92}$$

The integrand in (4.92) goes to zero as $|\lambda|$ goes to infinity in the left-half plane. Hence we can evaluate this integral by forming a closed contour over the left-half plane, enclosing all poles of $F(\lambda)$:

$$\boxed{F_T(z, \tau) = z^{-l}\sum \text{residues of } \frac{e^{\delta T\lambda}F(\lambda)}{1 - e^{\lambda T}z^{-1}} \text{ at poles of } F(\lambda)} \tag{4.93}$$

The z transform $F_T(z, \tau)$ is thus a more general form of the z transformation $F_T(z)$ and is applicable whenever $\{f(t)\}$ is defined for all t in a given interval. Note that if $l = 1$ and δ varies continuously from 0 to 1, $F_T(z, \tau)$ will cover the samples of $\{f(t)\}$ for *all* t. Thus, although $F_T(z)$ can describe $\{f(t)\}$ only at the instants $t = kT$, $F_T(z, \tau)$, where now $\tau = T - \delta T$ and δ is permitted to vary continuously from 0 to 1, can be regarded as a complete and unambiguous representation of $\{f(t)\}$.

If $\{f(t)\}$ has a discontinuity at $t = k_0 T$, letting $l = k_0$ and $\delta = 0$ in (4.93) will give an $F_T(z)$ which places at $t = k_0 T$ the value of $\{f(t)\}$ at $t = (k_0 T)_+$. However, letting $l = k_0 + 1$ and $\delta = 1$ yields an $F_T(z)$ which places at $t = k_0 T$ the value of $\{f(t)\}$ at $t = (k_0 T)_-$. This is consistent with the convention adopted in connection with the derivation of (4.16) in Section 4.2.

Example

Given

$$f(t) = a^t, \qquad t \geq 0$$
$$= 0, \qquad t < 0$$

find

$$\mathcal{Z}[f(t - \tau)]$$

Let

$$\tau = lT - \delta T, \qquad 0 \leq \delta \leq 1$$

Using (4.93), we have

$$F(\lambda) = \frac{1}{\lambda - \log a}, \qquad \text{pole at } \lambda = \log a$$

$$F_T(z, \tau) = z^{-l} \frac{e^{\delta T \lambda}}{1 - e^{\lambda T} z^{-1}} \bigg|_{\lambda = \log a}$$

$$= \frac{z^{-l} a^{\delta T}}{1 - a^T z^{-1}}$$

For $l = 0$, $\delta = 0$,

$$F_T(z, \tau) = \frac{1}{1 - a^T z^{-1}}$$

while for $l = 1$, $\delta = 1$,

$$F_T(z, \tau) = \frac{a^T z^{-1}}{1 - a^T z^{-1}}$$

The second transform is identical with the first except that the second one gives $f(0) = 0$, that is, the value of $\{f(t)\}$ at 0_-.

The z transform of (4.93) is based on a mathematical model in which sampling occurs at $t = kT$, $k = 0, 1, 2, \ldots$, and in which the function $\{f(t)\}$ is shifted (i.e., delayed) in time by an amount τ. We write $\{f(t - \tau)\}$, where $\tau = lT - \delta T$, l is a positive integer, and $0 \leq \delta \leq 1$. Sometimes we encounter a situation where the sampling, though uniformly spaced, does not occur at $t = kT$ but at some shifted instants $t = kT + t_0$, where $t_0 = bT + \delta T$, b is a positive integer, and $0 \leq \delta < 1$. This situation is illustrated in Fig. 4-9.

We define a new variable $t' = t - \delta T$. For the new variable, $f(t)$ becomes $f(t' + \delta T)$, $t' \geq -\delta T$. Sampling occurs at $t' = kT$ for $k \geq b$.

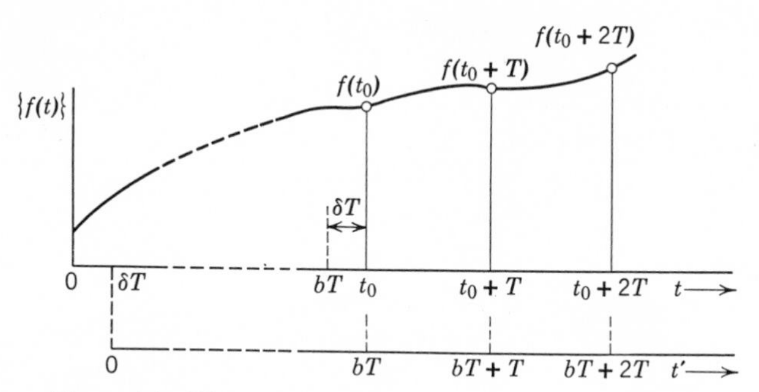

Figure 4-9 Illustration of sampling at instants $t = kT + t_0$.

We may now apply (4.93), letting $l = 0$. This yields the z transform of $f(t' + \delta T)$ for $t' = kT$ and $k \geq 0$, that is, $F_T(z, \tau)$ where $\tau = -\delta T$. To delete the samples for $k < b$, we write

$$\mathcal{Z}[f(t' + \delta T)] = F_T(z, \tau) - \sum_{k=0}^{b-1} f(kT + \delta T)z^{-k} \tag{4.94}$$

where $\mathcal{Z}[f(t' + \delta T)]$ is the z transform of the sequence obtained by sampling $\{f(t' + \delta T)\}$ at instants $t' = kT$ for $k \geq b$.

4.10 MULTIDIMENSIONAL SAMPLING

The process of sampling a function f defined for the continuous variable t can be extended to functions of two or more independent continuous variables. Such functions are encountered in connection with problems in heat flow, transmission lines, and statistics.

By analogy with (3.103), we may define a two-dimensional delta-function series

$$\boxed{\delta(t, x; T, X) = \sum_{k=0}^{\infty} \sum_{m=0}^{\infty} \delta(t - kT, x - mX)} \tag{4.95}$$

where the two-dimensional delta function is defined by means of the integral

$$\int_{-\infty}^{\infty} dt \int_{-\infty}^{\infty} dx\, \delta(t - kT, x - mX) f(t, x) = f(kT, mX) \tag{4.96}$$

The following symbolism is clearly equivalent:

$$\delta(t - kT, x - mX) = \delta(t - kT)\, \delta(x - mX) \tag{4.97}$$

The Laplace transform of $\{\delta(t, x; T, X)\}$ is given by

$$\Delta_{T,X}(s_1, s_2) = \sum_{k=0}^{\infty} \sum_{m=0}^{\infty} e^{-s_1 Tk} e^{-s_2 Xm} \tag{4.98}$$

$$= \frac{1}{(1 - e^{-s_1 T})(1 - e^{-s_2 X})} \tag{4.99}$$

where s_1 and s_2 are two distinct complex variables and one defines two abscissae of absolute convergence, σ_{at} and σ_{ax}.

The rules governing the existence of the two-dimensional Laplace transform $F(s_1, s_2)$ for a function $\{f(t, x)\}$ are the same as in the one-dimensional case (see Section 3.15). If a multidimensional function does not vanish when one or more of its variables are negative, we may employ the concept of the generating function in the same manner as for functions of one variable.

Thus we may write for a two-dimensional Laplace transform

$$\boxed{\mathscr{L}[f(t, x)] = F(s_1, s_2) = \int_{0}^{\infty} dt \int_{0}^{\infty} dx\, f(t, x) e^{-(s_1 t + s_2 x)}} \tag{4.100}$$

$$\boxed{\mathscr{L}^{-1}[F(s_1, s_2)] = f(t, x) = \left(\frac{1}{2\pi j}\right)^2 \int_{c_1 - j\infty}^{c_1 + j\infty} ds_1 \int_{c_2 - j\infty}^{c_2 + j\infty} ds_2\, F(s_1, s_2) e^{(s_1 t + s_2 x)}} \tag{4.101}$$

where $c_1 > \sigma_{at}$ and $c_2 > \sigma_{ax}$.

Utilizing the same concepts employed in the derivation of (4.8), we write

$$\mathscr{L}[f(t, x; T, X)] = \mathscr{L}[f(t, x)\, \delta(t, x; T, X)] \tag{4.102}$$

and, by analogy with (4.5),

$$\mathscr{L}[f_1(t, x) f_2(t, x)]$$

$$= \left(\frac{1}{2\pi j}\right)^2 \int_{c_1 - j\infty}^{c_1 + j\infty} d\lambda_1 \int_{c_2 - j\infty}^{c_2 + j\infty} d\lambda_2\, F_1(\lambda_1, \lambda_2) F_2(s_1 - \lambda_1, s_2 - \lambda_2) \tag{4.103}$$

Let $\{f_2(t, x)\} = \{\delta(t, x; T, X)\}$ and substitute for $F_2(s_1 - \lambda_1, s_2 - \lambda_2)$ from (4.99). Then

$$\mathscr{L}[f(kT, mX)] = \left(\frac{1}{2\pi j}\right)^2 \int_{c_1 - j\infty}^{c_1 + j\infty} d\lambda_1 \int_{c_2 - j\infty}^{c_2 + j\infty} d\lambda_2 \frac{F_1(\lambda_1, \lambda_2)}{(1 - e^{T(\lambda_1 - s_1)})(1 - e^{X(\lambda_2 - s_2)})} \tag{4.104}$$

where now $\sigma_{a1} < c_1 < \sigma_1 - \sigma_{a3}$ and $\sigma_{a2} < c_2 < \sigma_2 - \sigma_{a4}$, σ_{a1}, σ_{a2}, σ_{a3}, and σ_{a4} being the abscissae of absolute convergence for $F_1(\lambda_1, x)$, $F_2(t, \lambda_2)$, $\Delta_T(\lambda_1, x)$, and $\Delta_X(t, \lambda_2)$, respectively. Note that for the two-dimensional delta function $\sigma_{a3} = \sigma_{a4} = 0$.

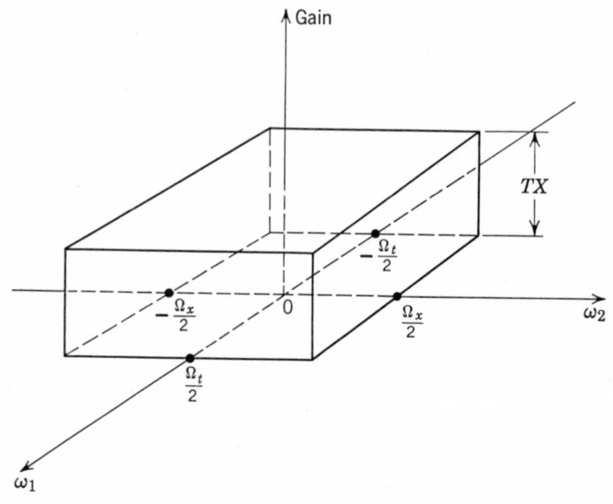

Figure 4-10 Two-dimensional ideal bandpass filter.

Thus, as in the case of (4.8), the integral (4.104) may be evaluated over either the right-half or left-half λ planes. The former leads to the expression

$$\mathscr{L}[f(t, x; T, X)] = \frac{1}{TX} \sum_{\nu=-\infty}^{\infty} \sum_{l=-\infty}^{\infty} F(s_1 + j\nu\Omega_t, s_2 + jl\Omega_x) \tag{4.105}$$

where $\Omega_t = 2\pi/T$ and $\Omega_x = 2\pi/X$. Evaluation over the left-half plane gives

$$F_{T,X}(z_1, z_2) = \sum_{k=0}^{\infty} \sum_{m=0}^{\infty} f(k, m) z_1^{-k} z_2^{-m}, \qquad z_1 > R_{c1}, z_2 > R_{c2} \tag{4.106}$$

where $z_1 = e^{sT}$ and $z_2 = e^{sX}$, and R_{c1} and R_{c2} are the radii of convergence in the z_1 and z_2 planes, respectively. The two expressions (4.105) and (4.106) are, of course, equivalent.

Note from (4.105) that the sampling theorem applies to both frequency variables of $F(s_1, s_2)$. If $F(\omega_1, \omega_2)$ is bandlimited such that $|F(\omega_1, \omega_2)| = 0$ for $|\omega_1| > \Omega_t/2$ and for $|\omega_2| > \Omega_x/2$, the function $\{f(t, x)\}$ can be uniquely determined from the samples $f(kT, mX)$. We need merely pass the samples through a two-dimensional filter of gain TX for $-\Omega_t/2 < \omega_1 < \Omega_t/2$ and $-\Omega_x/2 < \omega_2 < \Omega_x/2$, and of zero gain elsewhere, as shown in Fig. 4-10. This filter has the two-dimensional weighting function

$$w(t, x) = \frac{\sin \Omega_x x}{\Omega_x x} \frac{\sin \Omega_t t}{\Omega_t t} \tag{4.107}$$

The inversion integral for the two-dimensional z transform is found in the same manner as (3.54):

$$\boxed{f(k, m) = \left(\frac{1}{2\pi j}\right)^2 \oint_{\Gamma_1} dz_1 \oint_{\Gamma_2} dz_2\, z_1^{k-1} z_2^{m-1} F(z_1, z_2)} \tag{4.108}$$

The inversion integral may be evaluated by means of the residue method, subject to the same restrictions that were described in Section 3.8 in connection with the evaluation of (3.54).

REFERENCES

Balakrishnan, A. V., "On the Problem of Time Jitter in Sampling," *IRE Trans. Inform. Theory*, vol. IT-8, no. 3, pp. 226–236, April 1962.

Bergen, A. R., *The Synthesis of Optimum Random Sampling Systems*, Tech. Report T-T/133, Electronics Research Laboratory, Columbia University, New York, 1957.

Beutler, F. J., "Sampling Theorems and Bases in a Hilbert Space," *Inform. Control*, vol. 4, no. 2–3, pp. 97–117, September 1961.

Childers, D. G., "Study and Experimental Investigation on Sampling Rate and Aliasing in Time-Division Telemetry Systems," *IRE Trans. Space Electron. Telemetry*, vol. SET-8, no. 4, pp. 267–283, December 1962.

Fogel, L. J., "A Note on the Sampling Theorem," *IRE Trans. Inform. Theory*, vol. IT-1, pp. 47–48, March 1955.

Helms, H. D., *Generalizations of the Sampling Theorem and Error Calculations*, Ph.D. dissertation, Princeton University, January 1961.

Helms, H. D., and J. B. Thomas, "Truncation Error of Sampling-Theorem Expansions," *Proc. IRE*, vol. 50, no. 2, pp. 179–184, February 1962.

Jagerman, D. L., and L. J. Fogel, "Some General Aspects of the Sampling Theorem," *IRE Trans. Inform. Theory*, vol. IT-2, no. 4, pp. 139–146, December 1956.

Jury, E. I., *Sampled-Data Control Systems*, John Wiley and Sons, New York, 1958.
Landau, H. S. and H. O. Pollak, "Prolate Spheroidal Wave Functions, Fourier Analysis and Uncertainty—II," *Bell System Tech. J.*, vol. 40, pp. 65–84, January 1961.
Landau, H. S. and H. O. Pollak, "Prolate Spheroidal Wave Functions, Fourier Analysis and Uncertainty—III," *Bell System Tech. J.*, vol. 41, pp. 1295–1336, July 1962.
Landgrebe, D. A. and G. R. Cooper, "Two-Dimensional Signal Representation Using Prolate Spheroidal Functions," *IEEE Trans. Commun. Electron.*, vol. 82, no. 65, pp. 30–40, March 1963.
Linden, D. A., "A Discussion of Sampling Theorems," *Proc. IRE*, vol. 47, no. 7, pp. 1219–1226, July 1959.
Linden, D. A. and N. M. Abramson, "A Generalization of the Sampling Theorem," *Inform. Control*, vol. 3, no. 1, pp. 26–31, March 1960.
Lorens, C. S., "Recovery of Randomly Sampled Time Sequences," *IRE Trans. Commun. Systems*, vol. CS-10, no. 2, pp. 214–216, June 1962.
Nyquist, H., "Certain Topics in Telegraph Transmission Theory," *Trans AIEE*, vol. 47, pp. 617–644, April 1928.
Oliver, B. M., J. R. Pierce, and C. E. Shannon, "The Philosophy of PCM," *Proc. IRE*, vol. 36, no. 11, pp. 1324–1331, November 1948.
Petersen, Daniel P., and D. Middleton, "Sampling and Reconstruction of Wave-Number-Limited Functions in *N*-Dimensional Euclidean Space," *Inform. Control*, vol. 5, no. 4, pp. 279–323, December 1962.
Shannon, C. E., "Communications in the Presence of Noise," *Proc. IRE*, vol. 37, no. 1, pp. 10–21, January 1949.
Slepian, D. and H. O. Pollak, "Prolate Spheroidal Wave Functions, Fourier Analysis and Uncertainty—I," *Bell System Tech. J.*, vol. 40, pp. 43–64, January 1961.
Whittaker, E. T., "On the Functions Which are Represented by the Expansions on the Interpolation Theory," *Proc. Royal Soc. Edinburgh*, vol. 35, pp. 181–194, 1915.
Woodward, P. M., *Probability and Information Theory with Applications to Radar*, McGraw-Hill Book Co., New York, 1953.
Yen, J. L., "On Non-uniform Sampling of Bandwidth-Limited Signals," *IRE Trans. Circuit Theory*, vol. CT-3, pp. 251–257, December 1956.
Yen, J. L., "On the Synthesis of Line and Infinite Strip Sources," *IRE Trans. Antennas Propagation*, vol. AP-5, pp. 40–46, January 1957.

CHAPTER 5

Polynomial Interpolation and Extrapolation

5.1 INTRODUCTION

A problem of major importance in the study of discrete-time systems is that of generating a continuous-time function $\{\varphi(t)\}$ from given discrete data $\{f(kT)\}$ such that $\varphi(kT) = f(kT)$. This problem is quite different from the one discussed in Chapter 4 in connection with the sampling theorem. There we showed that a function f whose Fourier transform vanishes for $|\omega| > \Omega/2$ can be uniquely reconstructed from its samples $f(kT)$, where k ranges over all integers and $\Omega T = 2\pi$. When selecting signals to represent characteristics of physical phenomena, we can never assume that the signals be bandlimited. Moreover, we cannot even assume that the signals be deterministic for all time; over a *long* time span, probabilistic factors must be taken into account. Finally, if reconstruction is to take place in *real* time,[1] only a nonanticipatory system can be employed. Hence in this chapter we shall disregard the question of *reconstruction*. Instead we shall merely try to find continuous-time functions that will *smoothly* interpolate over (or extrapolate from) a finite set of samples of an otherwise unknown function.

5.2 POLYNOMIAL INTERPOLATION

Given a set of $m + 1$ equidistant data values $f(kT)$, $k = j, j - 1, \ldots, j - m$, there is one and only one polynomial $\{\varphi(t)\}$ of degree m such that $\varphi(kT) = f(kT)$ for $j - m \leq k \leq j$. This polynomial will have $m + 1$ coefficients, all of which may be determined by solving the $m + 1$ simultaneous equations $\varphi(jT) = f(jT)$, $\varphi(jT - T) = f(jT - T)$, etc. The polynomial

$$\varphi(t) = a_0 + a_1 t + \cdots + a_m t^m \tag{5.1}$$

is unique; for if it were not, a polynomial $\varphi_1(t)$ of degree m would exist

[1] That is, in true clock time.

such that $\varphi_1(kT) = f(kT)$ for $j - m \leq k \leq j$, and the difference $\varphi(t) - \varphi_1(t)$ would be another polynomial of degree m such that it would have $m + 1$ zeros at $t = jT$. Since, however, a polynomial of degree m can have only m zeros, the only way for the difference polynomial to vanish at $m + 1$ points is for it to vanish everywhere, that is, $\varphi(t) \equiv \varphi_1(t)$. It follows, therefore, that if the values of $f(kT)$ are obtained by sampling a polynomial function $\{f(t)\}$ of degree m, the function $\{f(t)\}$ can be reconstructed unambiguously by solving $m + 1$ simultaneous equations $\varphi(kT) = f(kT)$. The resulting polynomial $\{\varphi(t)\}$ will be precisely equal to $\{f(t)\}$. The case where it is known *a priori* that $\{f(t)\}$ is a polynomial of degree m is a relatively trivial one. Of much greater interest is the case where a function[2] $\{f(t)\}$ is known to be such that it can be *approximated* to a specified precision by a polynomial of degree m in a given interval. The approximating polynomial $\{\varphi(t)\}$ can then be computed from a knowledge of $m + 1$ values of $f(kT)$. The approximation is in general better in the interior of the range $j - m \leq k \leq j$ than near or beyond either end.

If the process of determining $\{\varphi(t)\}$ occurs in a physical system (i.e., in "real" time), it usually takes the form of *extrapolation*. In that case the polynomial $\{\varphi(t)\}$ will be based on $m + 1$ past values of $f(kT)$ and the interval of approximation will be beyond the most recent value of $f(kT)$.

5.3 LAGRANGIAN POLYNOMIALS

The procedure for fitting polynomials to a set of discrete data is greatly facilitated by the use of the so-called Lagrangian polynomials. Let us assume that we are given $f(kT)$ for $m + 1$ successive values of k, $j - m \leq k \leq j$. We wish to determine a polynomial $\{\varphi(t)\}$ of degree m such that $\varphi(kT) = f(kT)$ for all $j - m \leq k \leq j$. Using (5.1), we have

$$
\begin{aligned}
f(jT) &= a_0 + a_1 jT + a_2 j^2 T^2 + \cdots + a_m j^m T^m \\
f(jT - T) &= a_0 + a_1(j-1)T + a_2(j-1)^2 T^2 + \cdots + a_m(j-1)^m T^m \\
&\;\;\vdots \\
f(jT - mT) &= a_0 + a_1(j-m)T + a_2(j-m)^2 T^2 + \cdots + a_m(j-m)^m T^m
\end{aligned}
\tag{5.2}
$$

The $m + 1$ simultaneous equations in (5.2) together with (5.1) for $t \neq jT$ form a set of $m + 2$ equations in the $m + 1$ unknown constants of (5.1).

[2] The function may be deterministic or random.

For this to represent a valid system of simultaneous equations, the determinant of the coefficients of the $m + 2$ equations must vanish. We write

$$\begin{vmatrix} \varphi(t) & 1 & t & t^2 & \dots & t^m \\ f(jT) & 1 & jT & j^2T^2 & \dots & j^mT^m \\ f(jT-T) & 1 & (j-1)T & (j-1)^2T^2 & \dots & (j-1)^mT^m \\ \cdot & & & & & \cdot \\ \cdot & & & & & \cdot \\ \cdot & & & & & \cdot \\ f(jT-mT) & 1 & (j-m)T & (j-m)^2T^2 & \dots & (j-m)^mT^m \end{vmatrix} = 0 \quad (5.3)$$

We can solve (5.3) for $\varphi(t)$ and obtain an expression which gives $\varphi(t)$ as a polynomial in t with coefficients formed from the $m + 1$ values of $f(kT)$. Thus, for $m = 1$,

$$\varphi(t) = f(jT)\left[\frac{t-(j-1)T}{T}\right] + f(jT-T)\left[\frac{t-jT}{-T}\right] \quad (5.4)$$

For $m = 2$,

$$\begin{aligned} \varphi(t) = f(jT) &\left\{\frac{[t-(j-1)T][t-(j-2)T]}{2T^2}\right\} \\ &+ f(jT-T)\left\{\frac{[t-jT][t-(j-2)T]}{-T^2}\right\} \\ &+ f(jT-2T)\left\{\frac{[t-jT][t-(j-1)T]}{2T^2}\right\} \end{aligned} \quad (5.5)$$

For any m,

$$\varphi(t) = \sum_{i=0}^{m}\left[\frac{\prod\limits_{\substack{l=0\\ l\neq i}}^{m}[t-(j-l)T]}{\prod\limits_{\substack{l=0\\ l\neq i}}^{m}[lT-iT]}\right] f(jT-iT) \quad (5.6)$$

Equation (5.6) can be put into simpler form if we let

$$B_m(t) = [t-jT][t-(j-1)T][t-(j-2)T]\dots[t-(j-m)T] \quad (5.7)$$

and define as a *Lagrangian polynomial*,

$$\boxed{L_{m,i}(t) = \frac{B_m(t)}{[t-(j-i)T]B_m'(jT-iT)}} \quad (5.8)$$

The approximation function $\{\varphi(t)\}$ is then given by

$$\boxed{\varphi(t) = \sum_{i=0}^{m} L_{m,i}(t) f(jT - iT)} \tag{5.9}$$

The quantity $B_m'(jT - iT)$ is to be interpreted as

$$\lim_{t \to jT - iT} \frac{d}{dt} B_m(t)$$

We observe that the polynomial $\varphi(t)$ of degree m is expressed in (5.9) as a sum of Lagrangian polynomials of degree m, each multiplied by the appropriate value of $f(kT)$, $j - m \leq k \leq j$. The Lagrangian polynomials $L_{m,i}(t)$ have the property that

$$L_{m,i}(iT) = 1 \tag{5.10}$$

$$L_{m,i}(kT) = 0 \quad \text{for } k \neq i \tag{5.11}$$

$$\sum_{i=0}^{m} L_{m,i}(t) = 1 \tag{5.12}$$

The significance of the Lagrangian polynomials can be seen from (5.9). The various $L_{m,i}(t)$ need be computed only once and may then be stored in tabular form. To determine the function $\{\varphi(t)\}$ for a particular set of m successive values of $f(kT)$, we merely substitute in (5.9). It is not necessary to solve any simultaneous equations!

As an illustration, assume that we are given a set of successive values $f(kT)$ in "real" time and wish to find a function $\{\varphi(\tau)\}$, $0 \leq \tau < T$, which will permit us to extrapolate from the current value, $f(jT)$, to the future value, $f(jT + T)$. The values $f(kT)$ are known to follow closely a parabolic law.

Let $m = 2$ and $t = jT + \tau$ in (5.8), and we obtain

$$L_{2,0}(\tau) = \frac{(\tau + T)(\tau + 2T)}{2T^2} \tag{5.13}$$

$$L_{2,1}(\tau) = \frac{\tau(\tau + 2T)}{-T^2} \tag{5.14}$$

$$L_{2,2}(\tau) = \frac{\tau(\tau + T)}{2T^2} \tag{5.15}$$

Then

$$\varphi(\tau) = f(kT)L_{2,0}(\tau) + f(kT - T)L_{2,1}(\tau) + f(kT - 2T)L_{2,2}(\tau) \tag{5.16}$$

for $0 \leq \tau < T$.

We may represent the extrapolation process of (5.16) by means of a

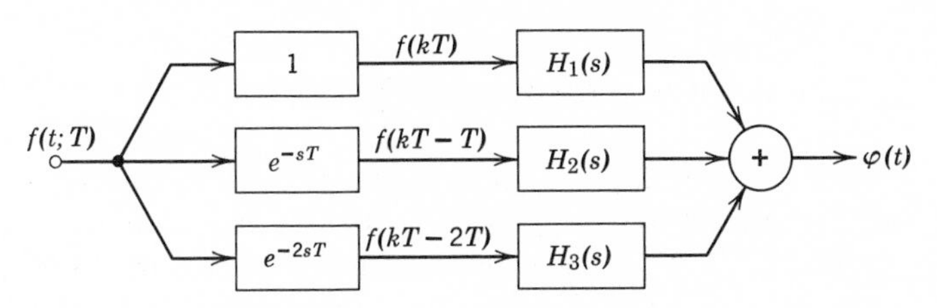

Figure 5-1 Block diagram of 2nd-degree polynomial extrapolation.

block diagram as shown in Fig. 5-1. The inputs to this system are the values of $f(kT)$, that is, the values of $\{f(t; T)\}$ at $t = kT$, and the output is $\varphi(t)$ where $t = kT + \tau$ and $0 \leq \tau < T$.

The transfer functions $H_1(s)$, $H_2(s)$, and $H_3(s)$ are respectively the Laplace transforms of the weighting functions $L_{2,0}(\tau)$, $L_{2,1}(\tau)$, and $L_{2,2}(\tau)$. We find

$$H_1(s) = \frac{1 - e^{-sT}}{T^2 s^3} + \frac{3 - 5e^{-sT}}{2Ts^2} + \frac{1 - 3e^{-sT}}{s} \tag{5.17}$$

$$H_2(s) = -\frac{2(1 - e^{-sT})}{T^2 s^3} - \frac{2(1 - 2e^{-sT})}{Ts^2} + \frac{3e^{-sT}}{s} \tag{5.18}$$

$$H_3(s) = \frac{1 - e^{-sT}}{T^2 s^3} + \frac{1 - 3e^{-sT}}{2Ts^2} - \frac{e^{-sT}}{s} \tag{5.19}$$

The overall transfer function for the entire system is given by

$$H(s) = \frac{\mathscr{L}[\varphi(t)]}{\mathscr{L}[f(t; T)]} = \frac{(2 + 3Ts + 2T^2 s^2)(1 - e^{-sT})^3}{2T^2 s^3} \tag{5.20}$$

The corresponding impulse response (i.e., weighting function) is shown in Fig. 5-2.

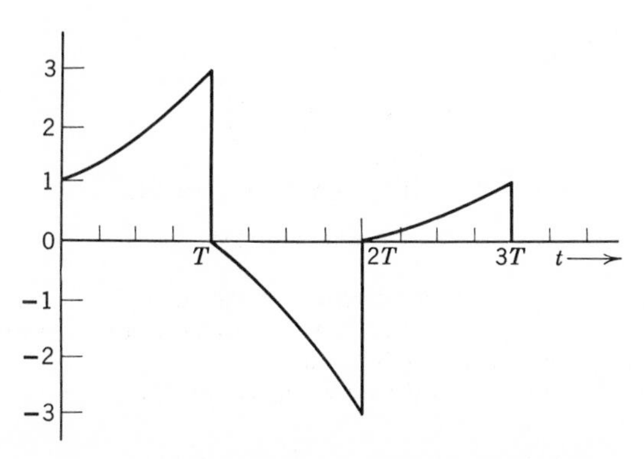

Figure 5-2 Unit impulse response of 2nd-degree polynomial extrapolation.

5.4 APPROXIMATION ERROR

We have shown that a polynomial $\{\varphi(t)\}$ of degree m can be obtained which will represent $\{f(t)\}$ without error for all t if we are given $m + 1$ values $f(kT)$, and it is known that $\{f(t)\}$ is of polynomial form and of degree equal to or less than m. However, if $\{f(t)\}$ is of degree greater than m, $\{\varphi(t)\}$ will be merely an approximation to $\{f(t)\}$, except, of course, that $\varphi(t) = f(t)$ at the $m + 1$ given values $f(kT)$. Although we cannot determine the actual magnitude of the error without knowing $\{f(t)\}$, we can frequently determine an upper bound on its value.

Consider the equation

$$q(t) = f(t) - \varphi(t) - cB_m(t) \tag{5.21}$$

where $B_m(t)$ is defined as in (5.7) and where c is selected such that $q(t) = 0$ for some value $t = t_0$. The choice of t_0 is arbitrary except that $jT - mT < t_0 < jT$ and $t_0 \neq iT$ for any integer i. Now the equation

$$q(t) = f(t) - \varphi(t) - cB_m(t) = 0 \tag{5.22}$$

will have $m + 2$ roots; that is, $t = jT, jT - T, \ldots, jT - mT$, and t_0. From Rolle's theorem,[3] the derivative $q'(t)$ will vanish at least $m + 1$ times in the interval $jT - mT < t < jT$. Similarly, the second derivative $q''(t)$ will vanish at least m times in this interval. Finally, the $(m + 1)$th derivative, $q^{(m+1)}(t)$, will vanish at least at one point $t = \xi$ in this interval. Thus

$$q^{(m+1)}(\xi) = 0 \tag{5.23}$$

Hence if we differentiate each term in (5.21) $m + 1$ times and let $t = \xi$, we shall have

$$0 = f^{(m+1)}(\xi) - \varphi^{(m+1)}(\xi) - cB_m^{(m+1)}(\xi) \tag{5.24}$$

Now, for $B_m(t)$ in (5.7), we can write[4]

$$B_m(t) = T^{m+1}\left(\frac{t}{T} - j\right)^{[m+1]} \tag{5.25}$$

Differentiating $m + 1$ times with respect to t, we find

$$B_m^{(m+1)}(t) = (m + 1)! \tag{5.26}$$

[3] "If $f(x)$ is single-valued and has a derivative throughout an interval $a \leq x \leq b$, and if both $f(a)$ and $f(b)$ are equal to zero, then there exists at least one value x_0, where $a < x_0 < b$, such that the derivative $f'(x_0) = 0$."

[4] We shall employ the notation for the ascending factorial function:

$$\left(\frac{t}{T} - j\right)^{[m+1]} = \left(\frac{t}{T} - j\right)\left(\frac{t}{T} - j + 1\right)\left(\frac{t}{T} - j + 2\right)\cdots\left(\frac{t}{T} - j + m\right).$$

Since $\{\varphi(t)\}$ is a polynomial of degree m,

$$\varphi^{(m+1)}(t) = 0 \tag{5.27}$$

for all t. Substituting (5.26) and (5.27) in (5.24) and solving for c gives

$$c = \frac{1}{(m+1)!} f^{(m+1)}(\xi) \tag{5.28}$$

and upon substitution in (5.21) we obtain for the approximation error $E(t) = f(t) - \varphi(t)$

$$\boxed{E(t) = \frac{B_m(t)}{(m+1)!} f^{(m+1)}(\xi)} \tag{5.29}$$

In practice we shall know neither the $(m+1)$th derivative of $\{f(t)\}$ nor the value ξ (which is a function of t). However, we may know that $f^{(m+1)}(t)$ does not exceed some upper bound M for $jT - mT \leq t \leq jT$. If we then make use of the fact that in this interval

$$|B_m(t)| < T^{m+1}(m+1)! \tag{5.30}$$

we obtain as an upper bound on the error[5]

$$\boxed{|E(t)| < MT^{m+1}} \tag{5.31}$$

for all $jT - mT \leq t \leq jT$, that is, for all cases where we *interpolate*.

It is easily verified that (5.30) applies even if we *extrapolate* over the interval $jT \leq t \leq jT + T$. As indicated previously, the error will, of course, normally be greater for extrapolation than for interpolation. The error bound is still given by (5.31), provided we choose M such that $M > \max f^{(m+1)}(t)$ in the interval $jT - mT \leq t \leq jT + T$. For extrapolation beyond $t = jT + T$, however, both (5.30) and (5.31) require modification.

5.5 NEWTON-GREGORY EXTRAPOLATION

When extrapolating $\{f(t)\}$ over an interval $jT \leq t < jT + T$ on the basis of m values of $f(kT)$, $j - m \leq k \leq j$, we are in effect attempting to predict future values $f(t)$ from past data. This suggests another approach to the problem of finding $\{\varphi(t)\}$. Consider the process represented by the

[5] Equation (5.30) gives a pessimistic value for the upper bound on $B_m(t)$. It is possible to determine the least upper bound precisely for a given m and T by referring to (5.7).

block diagram of Fig. 5-3. The input consists of the function $\{f(t)\}$ and the output is the predicted function $\{f(t+\tau)\}$ where τ is a parameter lying between 0 and T. If we let $\mathscr{L}[f(t)] = F(s)$, then $\mathscr{L}[f(t+\tau)]$ is given by $e^{s\tau}F(s)$; that is, $e^{s\tau}$ is the transfer function for perfect prediction over a time interval τ. This is of course not physically realizable; however, we may use a suitable approximation. We write

$$e^{s\tau} = [1 - (1 - e^{-sT})]^{-\tau/T} \tag{5.32}$$

$$= 1 + \frac{\tau}{T}(1 - e^{-sT}) + \frac{1}{2!}\frac{\tau}{T}\left(\frac{\tau}{T} + 1\right)(1 - e^{-sT})^2 + \cdots + \frac{1}{i!}\left(\frac{\tau}{T}\right)^{[i]}(1 - e^{-sT})^i + \ldots \tag{5.33}$$

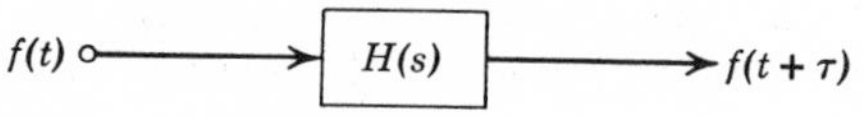

$f(t)$ → $H(s)$ → $f(t+\tau)$

Figure 5-3 Block diagram of extrapolation process.

where τ and T are fixed parameters. The expansion converges for $|1 - e^{-sT}| < 1$. Hence[6]

$$\mathscr{L}[f(t+\tau)] = \sum_{i=0}^{\infty}\frac{1}{i!}\left(\frac{\tau}{T}\right)^{[i]}(1 - e^{-sT})^i F(s) \tag{5.34}$$

Recognizing that

$$(1 - e^{-sT})^i F(s) = \mathscr{L}[\nabla^i f(t)] \tag{5.35}$$

where ∇ is the familiar backward-difference operator, we invert to the time domain and find

$$f(t+\tau) = \sum_{i=0}^{\infty}\frac{1}{i!}\left(\frac{\tau}{T}\right)^{[i]}\nabla^i f(t) \tag{5.36}$$

Since we are given $f(t)$ only at the discrete instants $t = kT$, we write

$$f(kT+\tau) = \sum_{i=0}^{\infty}\frac{1}{i!}\left(\frac{\tau}{T}\right)^{[i]}\nabla^i f(kT) \tag{5.37}$$

$$= f(kT) + \frac{\tau}{T}\nabla f(kT) + \frac{1}{2!}\left(\frac{\tau}{T}\right)\left(\frac{\tau}{T} + 1\right)\nabla^2 f(kT) + \ldots \tag{5.38}$$

Equation (5.38) is recognized as the familiar Newton-Gregory backward-difference formula. If $\{f(t)\}$ is a polynomial of degree m, that is, similar in form to (5.1), then $\nabla^i f(kT)$ in (5.37) will vanish for $i > m$ and the

[6] $\left(\frac{\tau}{T}\right)^{[i]} = \frac{\tau}{T}\left(\frac{\tau}{T} + 1\right)\left(\frac{\tau}{T} + 2\right)\ldots\left(\frac{\tau}{T} + i - 1\right)$

right-hand side of (5.38) will become an mth-degree polynomial in τ, identical to the polynomial $\{f(t)\}$ for $t = kT + \tau$. Again, since the polynomial representation is unique, such a polynomial will also be identical to the one obtained by the Lagrangian procedure in (5.9). Thus if in (5.9) we let $t = kT + \tau, j = k$, and $i = l$, we have the identity

$$\sum_{l=0}^{m} L_{m,l}(kT + \tau)f(kT - lT) \equiv \sum_{i=0}^{m} \frac{1}{i!}\left(\frac{\tau}{T}\right)^{[i]} \nabla^i f(kT) \tag{5.39}$$

for all τ and k.

Let us return to (5.37). In a typical case the nature of $\{f(t)\}$ is unknown and we must truncate the summation. Hence we write for $t = kT + \tau$

$$\boxed{\varphi(t) = \sum_{i=0}^{m} \frac{1}{i!}\left(\frac{\tau}{T}\right)^{[i]} \nabla^i f(kT)} \tag{5.40}$$

where $\{\varphi(t)\}$ is an mth degree polynomial approximation to $\{f(t)\}$; $m + 1$ successive values $f(kT)$ are required for the determination of the backward differences $\nabla^i f(kT)$.

The error $E(t) = f(t) - \varphi(t)$ in the extrapolation interval $kT \leq t \leq kT + T$ is, in accordance with (5.39), again given by (5.29). An upper bound on the error is given by (5.31) where M is again defined as

$$M = \max f^{(m+1)}(t) \tag{5.41}$$

for $jT - mT \leq t \leq jT + T$.

The signals employed to represent the time-varying characteristics of physical phenomena must always be permitted to exhibit random variations as well as to possess superimposed noise.[7] Hence high-order derivatives will always be present in such signals. Accordingly, it will be found that for a given spacing T the error (5.29) will initially decrease with increasing m but will then increase again as m increases beyond a certain value. In general, the greater the spacing T, the smaller the critical value of m beyond which the error increases.

Thus contrary to what (5.38) might lead us to believe, the approximation does not necessarily improve as we include more and more terms of the infinite series. The discrepancy is directly attributable to the fact that over progressively larger time intervals, $\{f(t)\}$ tends to become less and less deterministic in nature. In practice it is, therefore, necessary to restrict the use of (5.40) to small values of m. We shall try to interpret the approximations thus obtained.

[7] Any unwanted signal component is referred to as "noise."

5.6 ZERO-ORDER EXTRAPOLATION

If in (5.40) we let $m = 0$, we have

$$\varphi_0(t) = \varphi_0(kT + \tau) = f(kT) \tag{5.42}$$

that is, the value $\varphi_0(kT + \tau)$ is the same for all $0 \leq \tau < T$, and $\{\varphi_0(t)\}$ is thus of "staircase" form as shown in Fig. 5-4. Since only the zero-order term of (5.40) is used, this type of extrapolation is also referred to as *zero-order extrapolation* or a "zero-order hold."

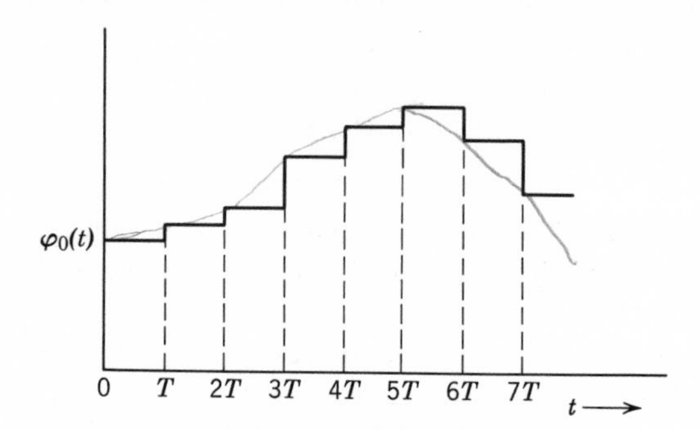

Figure 5-4 Staircase extrapolation.

If we regard the foregoing as a system with input $\{f(t; T)\}$ and output $\{\varphi_0(t)\}$, where $\varphi_0(kT + \tau) = f(kT)$, we can represent it by the transfer function

$$J_h{}^0(s) = \frac{1 - e^{-sT}}{s} \tag{5.43}$$

5.7 FIRST-ORDER EXTRAPOLATION

If in (5.40) we let $m = 1$, we obtain

$$\varphi_1(kT + \tau) = f(kT) + \frac{\tau \nabla f(kT)}{T} \tag{5.44}$$

where $\nabla f(kT) = f(kT) - f(kT - T)$. We can interpret this as giving $\varphi_1(t)$ in the interval $kT \leq t \leq kT + T$ in terms of a sum of the last value $f(kT)$ and a slope correction obtained by extending the slope of the straight line between the past two values $f(kT)$ and $f(kT - T)$. This is shown in

Fig. 5-5. During the interval $kT \leq t < kT + T$,

$$\varphi_1(t) = f(kT)\left(1 + \frac{\tau}{T}\right) - \frac{\tau}{T} f(kT - T) \tag{5.45}$$

and during the following interval

$$\varphi_1(t) = f(kT + T)\left(1 + \frac{\tau}{T}\right) - \frac{\tau}{T} f(kT) \tag{5 46}$$

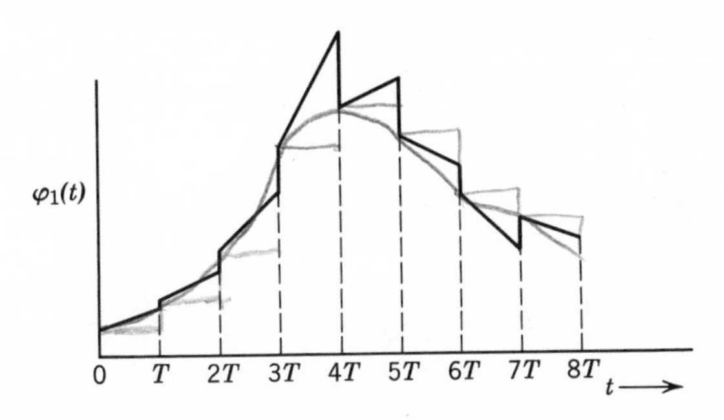

Figure 5-5 First-difference extrapolation.

The value $f(kT)$ does not affect the determination of $\varphi_1(t)$ in any other interval. Accordingly, the unit impulse response of a system performing first-difference extrapolation will be

$$j_h^1(t) = \left(1 + \frac{t}{T}\right)[v(t) - v(t - T)] - \frac{t - T}{T}[v(t - T) - v(t - 2T)] \tag{5.47}$$

where $v(t)$ is the familiar unit step function

$$\begin{aligned} v(t) &= 1 \text{ for } t \geq 0 \\ &= 0 \text{ for } t < 0 \end{aligned} \tag{5.48}$$

A sketch of $j_h^1(t)$ is given in Fig. 5-6. The corresponding transfer function is found to be

$$\boxed{J_h^1(s) = \frac{Ts + 1}{T}\left(\frac{1 - e^{-sT}}{s}\right)^2} \tag{5.49}$$

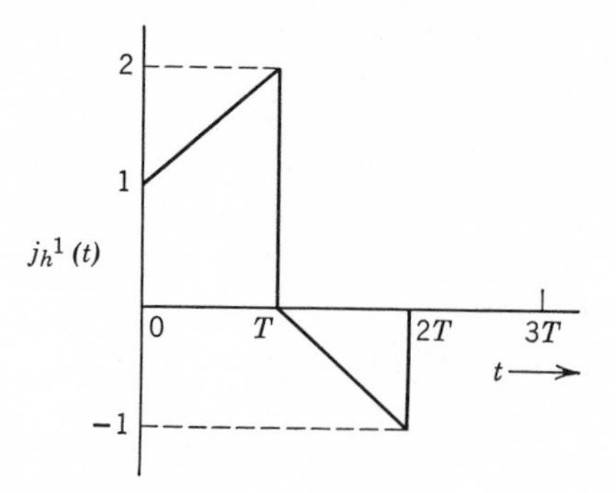

Figure 5-6 Unit impulse response of first-difference extrapolator.

5.8 SECOND-ORDER EXTRAPOLATION

For the case where $m = 2$ in (5.40) we obtain a second-difference extrapolator. In accord with the identity (5.39), this is identical with the extrapolation represented by (5.16) and (5.20); that is,

$$J_h^2(s) = \frac{(2 + 3Ts + 2T^2s^2)}{2T^2}\left(\frac{1 - e^{-sT}}{s}\right)^3 \tag{5.50}$$

The corresponding unit impulse response was shown in Fig. 5-2.

5.9 FREQUENCY-DOMAIN CHARACTERISTICS OF EXTRAPOLATION FUNCTIONS

It was noted in Chapter 4 that if a function $\{f(t)\}$ whose amplitude frequency spectrum $F(j\omega)$ is zero for $|\omega| > \Omega/2$ is sampled at a frequency Ω, the original function can always be recovered from the resulting samples $f(kT)$, $T = 2\pi/\Omega$. Thus given a function $f(kT)$ obtained from a band-limited $\{f(t)\}$, we need perform only the convolution summation (4.33)

$$f(t) = \sum_{k=-\infty}^{\infty} f(kT)\frac{\sin \Omega(t - kT)/2}{\Omega(t - kT)/2} \tag{4.33}$$

to recover $\{f(t)\}$. The function

$$w(t) = \frac{\sin \Omega t/2}{\Omega t/2} \tag{5.51}$$

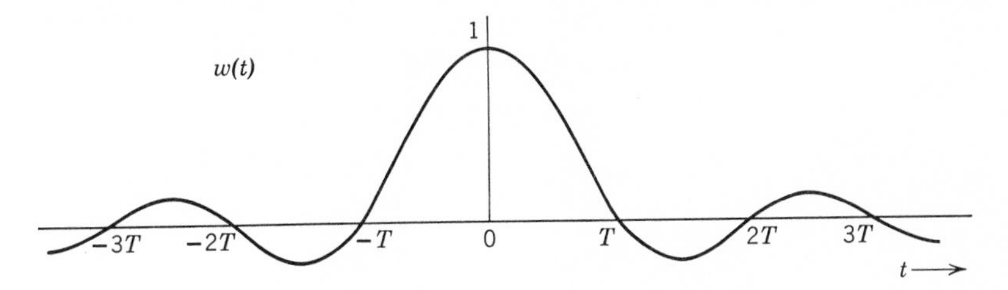

Figure 5-7 Cardinal interpolation function.

is known as the *cardinal interpolation function* and is shown in Fig. 5-7. We note that the function has unity value at $t = 0$ and is zero at all $t = kT$ where $T = 2\pi/\Omega$. A typical interpolation based on the use of $w(t)$ is shown in Fig. 5-8.

It is apparent from (4.33) that the cardinal interpolation function cannot be represented by a physically realizable process since its weighting function (5.51) ranges from $-\infty$ to $+\infty$. If a sufficiently long delay function is associated with it, it can be approximated; however, its attractive feature of providing perfect interpolation is then lost.

It is of interest to consider the amplitude-vs-frequency characteristics of the zero-order difference extrapolator. From (5.43),

$$J_h^0(j\omega) = \frac{1 - e^{-j\omega T}}{j\omega} \tag{5.52}$$

$$= T\frac{\sin \omega T/2}{\omega T/2} \angle -\frac{\omega T}{2} \tag{5.53}$$

The magnitude $|J_h^0(j\omega)|$ is plotted in Fig. 5-9. For comparison, the magnitude characteristic of the ideal filter $|J_h^i(j\omega)|$ (which led to the cardinal

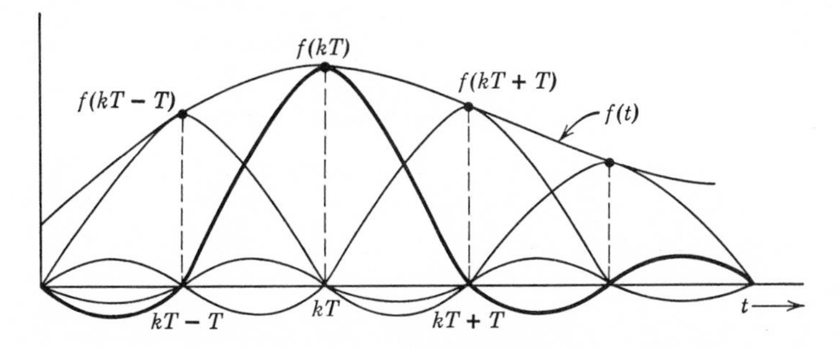

Figure 5-8 Illustration of cardinal function interpolation.

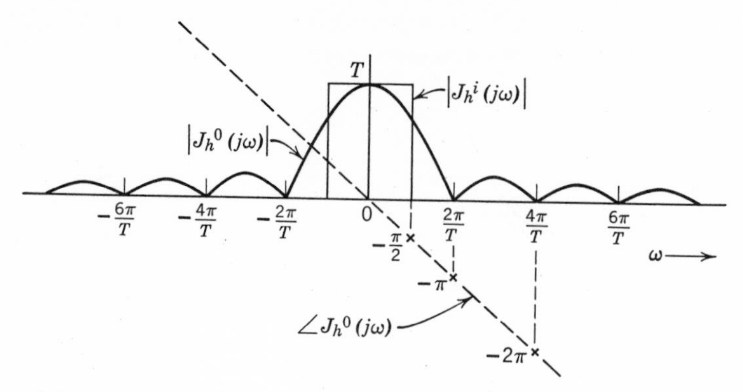

Figure 5-9 Magnitude and phase vs. frequency for zero-order extrapolation function $J_h^0(j\omega)$.

interpolation function) is shown superimposed. It is seen that the magnitude spectrum differs from that of an ideal filter, particularly so for $|\omega| > \pi/T$, and in addition there is a linear phase shift which equals $-\pi/2$ radians at $\omega = \pi/T$.

For the first-order extrapolation function $J_h{}^1(s)$ given by (5.49) we obtain, upon substituting $j\omega$ for s,

$$|J_h{}^1(j\omega)| = T\sqrt{1 + \omega^2T^2}\left(\frac{\sin \omega T/2}{\omega T/2}\right)^2 \tag{5.54}$$

$$\underline{/J_h{}^1(j\omega)} = -\omega T + \tan^{-1} \omega T \tag{5.55}$$

The magnitude (5.54) is shown plotted against frequency in Fig. 5-10. The phase angle from (5.55) is considerably greater than that for the zero-order extrapolator (5.53). Thus at $\omega = \pi/T$, the phase lag for $J_h{}^1(j\omega)$ is approximately -1.88 radians as compared to $-\pi/2$ radians for $J_h{}^0(j\omega)$. At $\omega = 2\pi/T$, $J_h{}^1(j\omega)$ has a phase lag of -4.87 radians while that for $J_h{}^0(j\omega)$ is only $-\pi$ radians. The greater phase lag associated with the first-order

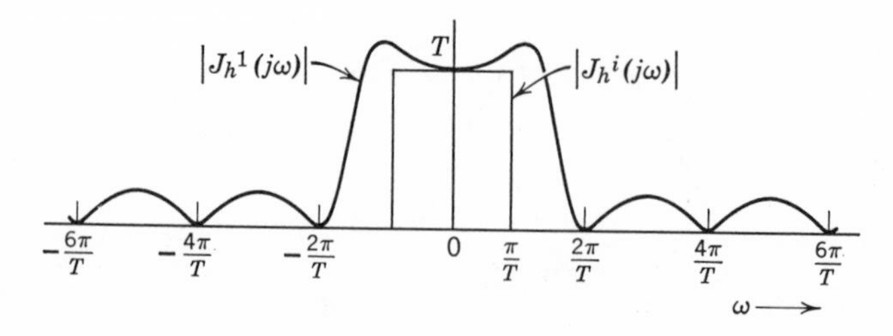

Figure 5-10 Magnitude vs. frequency for first-order extrapolation function $J_h^1(j\omega)$.

extrapolator is, of course, to be expected from the use of the additional past data. Higher-order extrapolators, such as $J_h^2(s)$, accordingly exhibit even greater phase lag.

REFERENCES

Hamming, R. W., *Numerical Methods for Scientists and Engineers*, McGraw-Hill Book Co., New York, 1962.

Herriot, John G., *Methods of Mathematical Analysis and Computation*, John Wiley and Sons, New York, 1963.

Hildebrand, F. B., *Introduction to Numerical Analysis*, McGraw-Hill Book Co., New York, 1956.

Kopal, Z., *Numerical Analysis* (2nd ed.), John Wiley and Sons, New York, 1961.

Lanczos, C., *Applied Analysis*, Prentice-Hall, Englewood Cliffs, N.J., 1956.

Porter, A., F. W. Stoneman, and D. F. Lawden, "A New Approach to the Design of Pulse-Monitored Servo Systems," *Proc. IEE*, vol. 97, part 2, pp. 507–610, London, October 1950.

Ragazzini, J. R. and G. Franklin, *Sampled-Data Control Systems*, McGraw-Hill Book Co., New York, 1958.

Salzer, John M., "Frequency Analysis of Digital Computers Operating in Real Time," *Proc. IRE*, vol. 42, no. 2, pp. 457–466, February 1954.

CHAPTER 6

Continuous-Time Systems with Discrete-Time Inputs

6.1 INTRODUCTION

In the previous chapters we have been concerned almost entirely with time sequences and linear discrete-time systems. We drew a careful distinction between continuous-time and discrete-time systems and regarded these as two separate system categories. In setting up a mathematical model for a dynamic phenomenon, we frequently find, however, that this simple dichotomy is too restrictive: many dynamic phenomena possess behavior characteristics, some of which can only be described in discrete time and others which are best described in continuous time. In such cases we desire a system model that consists of both discrete-time as well as continuous-time subsystems.

If a system consists of both discrete-time and continuous-time subsystems, the points of special interest are the interfaces between these subsystems. In going from a continuous-time to a discrete-time system, the output of the first must be *sampled* before it can serve as the input to the second. The sampling operation has been described in Chapter 4 and requires no further discussion. However, the other interface, where data from a discrete-time subsystem serves as the input to a continuous-time subsystem, requires some special analysis methods which we have not previously considered. We shall now investigate these methods, both for the case where the system is represented in terms of state equations as well as for the case where it is represented by means of transfer functions.

6.2 TRANSITION EQUATION OF A CONTINUOUS-TIME SYSTEM

Before proceeding further it is necessary to derive the transition equation for a continuous-time system. The derivation is analogous to that used in Section 2.4 for the transition equation of a discrete-time system. From

(1.29) and (1.30) we have, for a linear, stationary, continuous-time system

$$\frac{d}{dt}\mathbf{x}(t) = \mathbf{A}\mathbf{x}(t) + \mathbf{B}\mathbf{u}(t) \tag{6.1}$$

$$\mathbf{y}(t) = \mathbf{C}\mathbf{x}(t) \tag{6.2}$$

If (6.1) is expressed in terms of the Laplace transform, we have

$$s\mathbf{X}(s) - \mathbf{x}(0) = \mathbf{A}\mathbf{X}(s) + \mathbf{B}\mathbf{U}(s) \tag{6.3}$$

Solving for $\mathbf{X}(s)$, we obtain

$$\mathbf{X}(s) = [s\mathbf{I} - \mathbf{A}]^{-1}\mathbf{x}(0) + [s\mathbf{I} - \mathbf{A}]^{-1}\mathbf{B}\mathbf{U}(s) \tag{6.4}$$

Inverting to time domain then yields

$$\mathbf{x}(t) = e^{\mathbf{A}t}\mathbf{x}(0) + \int_0^{t^-} e^{\mathbf{A}(t-\tau)}\mathbf{B}\mathbf{u}(\tau)\,d\tau \tag{6.5}$$

where

$$e^{\mathbf{A}t} = \sum_{j=0}^{\infty} \frac{\mathbf{A}^j t^j}{j!} \tag{6.6}$$

and the integral on the right is the convolution of $e^{\mathbf{A}t}$ and $\mathbf{B}\mathbf{u}(t)$. Following the time convention introduced in Section 2.2, we shall not permit the input at time t, $\mathbf{u}(t)$, to affect the state at time t, $\mathbf{x}(t)$. For this reason the upper limit of the convolution integral in (6.5) is t^- rather than t.

If we let

$$\mathbf{\Phi}^c(t) = e^{\mathbf{A}t} \tag{6.7}$$

we can write

$$\mathbf{x}(t) = \mathbf{\Phi}^c(t)\mathbf{x}(0) + \int_0^{t^-} \mathbf{\Phi}^c(t-\tau)\mathbf{B}\mathbf{u}(\tau)\,d\tau \tag{6.8}$$

The matrix $\mathbf{\Phi}^c(t)$ is known as the *fundamental matrix* of the stationary continuous-time system.[1] It is the continuous-time counterpart of the discrete-time fundamental matrix defined in Section 2.4. It is readily seen that $\mathbf{\Phi}^c(t)$ possesses the following properties:

$$\mathbf{\Phi}^c(0) = \mathbf{I} \tag{6.9}$$

$$\mathbf{\Phi}^c(-t) = \mathbf{\Phi}^{c^{-1}}(t) \tag{6.10}$$

$$\mathbf{\Phi}^c(t + t_0) = \mathbf{\Phi}^c(t)\mathbf{\Phi}^c(t_0) \quad \text{for all } t \text{ and } t_0 \tag{6.11}$$

Equation (6.8) can be put into a slightly more general form. If we let $t = t_0$,

$$\mathbf{x}(t_0) = \mathbf{\Phi}^c(t_0)\mathbf{x}(0) + \int_0^{t_0^-} \mathbf{\Phi}^c(t_0-\tau)\mathbf{B}\mathbf{u}(\tau)\,d\tau \tag{6.12}$$

[1] The superscript c is used to distinguish this fundamental matrix from the fundamental matrix of a discrete-time-system.

After multiplying by $\mathbf{\Phi}^c(t - t_0)$ and rearranging, we have

$$\mathbf{\Phi}^c(t)\mathbf{x}(0) = \mathbf{\Phi}^c(t - t_0)\mathbf{x}(t_0) - \int_0^{t_0^-} \mathbf{\Phi}^c(t - \tau)\mathbf{Bu}(\tau)\, d\tau \tag{6.13}$$

Upon substitution in (6.8) we obtain the *transition equation*

$$\boxed{\mathbf{x}(t) = \mathbf{\Phi}^c(t - t_0)\mathbf{x}(t_0) + \int_{t_0}^{t^-} \mathbf{\Phi}^c(t - \tau)\mathbf{Bu}(\tau)\, d\tau} \tag{6.14}$$

The transition equation gives the state at time t, $\mathbf{x}(t)$, in terms of the state at time t_0 and the input $\mathbf{u}$ over the time interval $[t_0, t)$. As in the discrete-time case, the transition equation for a particular system is not unique but depends on the coordinate system selected in the state space.

Example

It is desired to determine the transition equation for a system that is described by the differential equation

$$\frac{d^2}{dt^2}y(t) + 5\frac{d}{dt}y(t) + 6y(t) = u(t)$$

Let

$$x_1(t) = y(t) \quad \text{and} \quad x_2(t) = \frac{d}{dt}y(t)$$

Then

$$\frac{d}{dt}x_1(t) = x_2(t)$$

$$\frac{d}{dt}x_2(t) = -6x_1(t) - 5x_2(t) + u(t)$$

and

$$\mathbf{A} = \begin{bmatrix} 0 & 1 \\ -6 & -5 \end{bmatrix}, \qquad \mathbf{B} = \begin{bmatrix} 0 \\ 1 \end{bmatrix}$$

From (6.4) and (6.8),

$$\mathscr{L}[\mathbf{\Phi}^c(t)] = [s\mathbf{I} - \mathbf{A}]^{-1} = \begin{bmatrix} s & -1 \\ 6 & s+5 \end{bmatrix}^{-1}$$

$$= \begin{bmatrix} \dfrac{s+5}{(s+2)(s+3)} & \dfrac{1}{(s+2)(s+3)} \\ \dfrac{-6}{(s+2)(s+3)} & \dfrac{s}{(s+2)(s+3)} \end{bmatrix}$$

Thus

$$\mathbf{\Phi}^c(t) = \begin{bmatrix} 3e^{-2t} - 2e^{-3t} & e^{-2t} - e^{-3t} \\ -6e^{-2t} + 6e^{-3t} & -2e^{-2t} + 3e^{-3t} \end{bmatrix}$$

Hence for the state-space coordinates selected, the transition equation of this system is

$$\mathbf{x}(t) = \begin{bmatrix} 3e^{-2(t-t_0)} - 2e^{-3(t-t_0)} & e^{-2(t-t_0)} - e^{-3(t-t_0)} \\ -6e^{-2(t-t_0)} + 6e^{-3(t-t_0)} & -2e^{-2(t-t_0)} + 3e^{-3(t-t_0)} \end{bmatrix}\mathbf{x}(t_0)$$

$$+ \int_{t_0}^{t^-} \begin{bmatrix} e^{-2(t-\tau)} - e^{-3(t-\tau)} \\ -2e^{-2(t-\tau)} + 3e^{-3(t-\tau)} \end{bmatrix} \mathbf{u}(\tau)\, d\tau$$

6.3 SYSTEM TRANSITION EQUATIONS FOR DISCRETE-TIME INPUTS

Let us now consider a system for which the input is represented by a modulated delta-function series of the form

$$\mathbf{u}(t;\, t_k) = \mathbf{u}(t)\sum_{k=0}^{\infty} \delta(t - t_k^+) \tag{6.15}$$

where the t_k satisfy the condition

$$t_{k+1} > t_k$$

but are otherwise arbitrary. If we replace $\mathbf{u}(\tau)$ in (6.14) by (6.15) and let $t = t_k + v$, where, for each interval $[t_k, t_{k+1}]$, $0 < v \leq t_{k+1} - t_k$, we obtain

$$\boxed{\mathbf{x}(t_k + v) = \mathbf{\Phi}^c(v)\mathbf{x}(t_k) + \mathbf{\Phi}^c(v)\mathbf{B}\mathbf{u}(t_k^+)} \tag{6.16}$$

Equation (6.16) is the transition equation for a continuous-time system for which the input consists of a sequence of impulses occurring at instants t_k^+, $k = 0, 1, 2, \ldots, \infty$.

If the delta-function series (6.15) is uniform, $t_{k+1} - t_k = T$, and we have

$$\boxed{\mathbf{x}(kT + v) = \mathbf{\Phi}^c(v)\mathbf{x}(kT) + \mathbf{\Phi}^c(v)\mathbf{B}\mathbf{u}(kT^+)} \tag{6.17}$$

where now $0 < v \leq T$. Finally, for $v = T$,

$$\boxed{\mathbf{x}(kT + T) = \mathbf{\Phi}^c(T)\mathbf{x}(kT) + \mathbf{\Phi}^c(T)\mathbf{B}\mathbf{u}(kT^+)} \tag{6.18}$$

The last equation is seen to be a *discrete-time state equation* for the continuous-time system. With input impulses applied to the system at $t = kT^+$, this equation gives the state of the system at $t = kT + T$ in terms of the state at $t = kT$ and the input at $t = kT^+$. Hence (6.18)—which is identical in form to (2.26)—can be solved by means of the discrete-time transition equation (2.29) for $\mathbf{x}(kT)$, the state at any specified time kT. Upon substituting $\mathbf{x}(kT)$ in (6.17), we obtain an expression for the state at any time $kT < t \leq kT + T$.

Equations (6.16) through (6.18) provide expressions for the state of a continuous-time system that is subjected to a uniform delta-function series input. There are, however, many systems in which the input is in the form of a "staircase" function; that is, the input changes from one constant value to another at successive instants t_k. We can say that in such systems the input is also in the form of a modulated delta-function series but is supplied to the system via a zero-order hold circuit.[2] We shall examine this case in some detail.

If in (6.14) we let $\mathbf{u}(\tau)$ be a constant, $\mathbf{u}(t_0)$, we obtain

$$\mathbf{x}(t) = \mathbf{\Phi}^c(t - t_0)\mathbf{x}(t_0) + \int_{t_0}^{t^-} \mathbf{\Phi}^c(t - \tau)\mathbf{B}\mathbf{u}(t_0)\, d\tau \tag{6.19}$$

Now let

$$\mathbf{\Lambda}^c(t - t_0) = \int_{t_0}^{t^-} \mathbf{\Phi}^c(t - \tau)\mathbf{B}\, d\tau \tag{6.20}$$

Then

$$\mathbf{x}(t) = \mathbf{\Phi}^c(t - t_0)\mathbf{x}(t_0) + \mathbf{\Lambda}^c(t - t_0)\mathbf{u}(t_0) \tag{6.21}$$

If the input to the system consists of a series of constant vectors $\mathbf{u}(t) = \mathbf{u}(t_k)$ for $t_k < t \leq t_{k+1}$, (6.21) becomes

$$\boxed{\mathbf{x}(t_k + v) = \mathbf{\Phi}^c(v)\mathbf{x}(t_k) + \mathbf{\Lambda}^c(v)\mathbf{u}(t_k)} \tag{6.22}$$

where $0 < v \leq t_{k+1} - t_k$ for each interval $[t_k, t_{k+1}]$.

If the steps change at uniformly spaced instants $t_k = kT$,

$$\boxed{\mathbf{x}(kT + v) = \mathbf{\Phi}^c(v)\mathbf{x}(kT) + \mathbf{\Lambda}^c(v)\mathbf{u}(kT)} \tag{6.23}$$

Hence, for $v = T$,

$$\boxed{\mathbf{x}(kT + T) = \mathbf{\Phi}^c(T)\mathbf{x}(kT) + \mathbf{\Lambda}^c(T)\mathbf{u}(kT)} \tag{6.24}$$

Equation (6.24) is a discrete-time state equation. It can be solved for $\mathbf{x}(kT)$ by means of (2.26). Together, (6.23) and (6.24) describe the state

[2] Cf. Section 5.6.

of a continuous-time system that is subjected to a staircase-type input which changes at uniformly spaced instants $t = kT$.

Example

A system described by the differential equation

$$\frac{d^2}{dt^2} y(t) + 5 \frac{d}{dt} y(t) + 6y(t) = u(t)$$

is subjected to a staircase input $u(t) = 4^k$ for $kT < t \leq kT + T$, $k = 0, 1, 2, \ldots, \infty$. We desire a state transition equation for the system.

Using $\mathbf{\Phi}^c(t)$ and $\mathbf{B}$ from the example at the end of Section 6.2, we have

$$\mathbf{\Phi}^c(t - \tau)\mathbf{B} = \begin{bmatrix} e^{-2(t-\tau)} - e^{-3(t-\tau)} \\ -2e^{-2(t-\tau)} + 3e^{-3(t-\tau)} \end{bmatrix}$$

Using (6.20), we obtain

$$\mathbf{\Lambda}^c(v) = \begin{bmatrix} \frac{1}{6} - \frac{1}{2}e^{-2v} + \frac{1}{3}e^{-3v} \\ e^{-2v} - e^{-3v} \end{bmatrix}$$

Hence from (6.24),

$$\mathbf{x}(kT + T) = \begin{bmatrix} 3e^{-2T} - 2e^{-3T} & e^{-2T} - e^{-3T} \\ -6e^{-2T} + 6e^{-3T} & -2e^{-2T} + 3e^{-3T} \end{bmatrix} \mathbf{x}(kT) + \begin{bmatrix} \frac{1}{6} - \frac{1}{2}e^{-2T} + \frac{1}{3}e^{-3T} \\ e^{-2T} - e^{-3T} \end{bmatrix} 4^k$$

This is of the form

$$\mathbf{x}(kT + T) = \mathbf{A}_1\mathbf{x}(kT) + \mathbf{B}_1\mathbf{u}(kT)$$

and can be solved for $\mathbf{x}(kT)$ by means of (2.29), yielding a result of the form

$$\mathbf{x}(kT) = \mathbf{A}_1{}^k\mathbf{x}(0) + \sum_{j=0}^{k-1} \mathbf{A}_1{}^j\mathbf{B}_1 4^{k-j-1}$$

[The technique of (3.48) may be used to determine $\mathbf{A}^k$]. Then, from (6.23),

$$\mathbf{x}(kT + v) = \begin{bmatrix} 3e^{-2v} - 2e^{-3v} & e^{-2v} - e^{-3v} \\ -6e^{-2v} + 6e^{-3v} & -2e^{-2v} + 3e^{-3v} \end{bmatrix} \mathbf{x}(kT) + \begin{bmatrix} \frac{1}{6} - \frac{1}{2}e^{-2v} + \frac{1}{3}e^{-3v} \\ e^{-2v} - e^{-3v} \end{bmatrix} 4^k$$

Observe that the actual determination of $\mathbf{x}(kT + v)$ is rather tedious—even in this very simple example. In a practical case a problem of this type must be solved on a digital computer.

6.4 SYSTEMS CHARACTERIZED BY TRANSFER FUNCTIONS

When dealing with relatively simple continuous-time systems, it is often preferable to characterize the systems by means of transfer functions rather than to use the state-space representation. We shall now consider the case where a discrete-time input (in the form of a modulated delta-function series) serves as the input to a continuous-time system described by a transfer function.

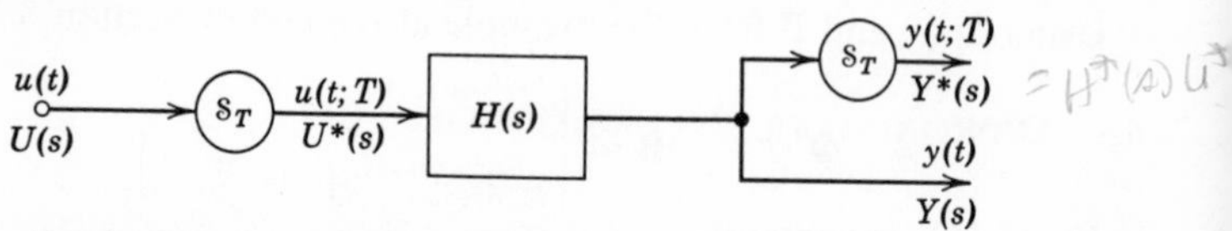

Figure 6-1 Continuous-time system with discrete-time input. *Note*: S_T represents a sampling operator which forms $\{f(t; T)\}$ from $\{f(t)\}$.

The system illustrated in Fig. 6-1 is represented by the transfer function $H(s)$. A signal $\{u(t)\}$, of Laplace transform $U(s)$, is sampled at $t = kT$, $k = 0, 1, 2, \ldots, \infty$ (as indicated by the sampling operator S_T), and a modulated delta-function series $\{u(t; T)\}$ is formed. The latter serves as the input to $H(s)$. The output of the system is represented by $\{y(t)\}$; the corresponding Laplace transform is $Y(s)$. Figure 6-1 also shows an output $\{y(t; T)\}$, obtained by sampling $\{y(t)\}$ at $t = kT$. The Laplace transforms of $\{u(t; T)\}$ and $\{y(t; T)\}$ will be denoted by $U^*(s)$ and $Y^*(s)$, respectively.

For any system, regardless of whether the input is continuous or discrete in time, if the Laplace transforms for the input and the system weighting function can be defined and there exists a common region of convergence between them, the Laplace transform of the output is simply given by the product of the input transform and the system transfer function. Thus in Fig. 6-1,

$$Y(s) = H(s)U^*(s) \tag{6.25}$$

Now from (4.21),

$$U^*(s) = \frac{1}{T}\sum_{p=-\infty}^{\infty} U(s + jp\Omega) \tag{6.26}$$

where $\Omega = 2\pi/T$. Hence

$$Y(s) = \frac{1}{T}\sum_{p=-\infty}^{\infty} H(s)U(s + jp\Omega) \tag{6.27}$$

If the alternate form of $U^*(s)$ is used, we have, from (4.15),

$$Y(s) = H(s)\sum_{k=0}^{\infty} u(kT)e^{-sTk} \tag{6.28}$$

The inversion of (6.27) or (6.28) is unfortunately not as simple as in the case of either a purely continuous-time or a purely discrete-time case. In the continuous-time case, $Y(s)$ is usually a rational function of s; in the discrete-time case, it is rational in $z = e^{sT}$. However, in the case of (6.25) it is neither rational in s nor in e^{sT}. Accordingly, a variety of special techniques must be employed to obtain $\{y(t)\}$ from $Y(s)$. Some of these give $\{y(t)\}$ precisely for all t. Some are approximate; that is, they describe $\{y(t)\}$ *approximately* for all t. Others are exact but incomplete; that is, they give $\{y(t)\}$ precisely only at certain selected instants of time.

6.5 APPROXIMATE INVERSION OF $Y(s)$

If in (6.27), $H(s)$ has a "low-pass" frequency characteristic, a fairly simple approximation to $Y(s)$ can be obtained by considering only a finite number of terms of the infinite summation.[3]

As an example, consider a system described by the transfer function $1/s(s + a)$. The system is subjected to an input whose Laplace transform is $V/(1 - e^{-bT}e^{-sT})$ (corresponding to a delta-function series Ve^{-bTk}, $k = 0, 1, 2, \ldots$). We obtain for the output $Y(s)$

$$Y(s) = \frac{V}{s(s + a)(1 - e^{-bT}e^{-sT})} \tag{6.29}$$

However, since $\mathscr{L}[e^{-bT}] = 1/(s + b)$, the input Laplace transform can also be written as[4]

$$\frac{1}{T}\sum_{l=-\infty}^{\infty} \frac{V}{s + b + jl\Omega}$$

Using only the terms for $l = 0, +1$, and -1, we have

$$Y(s) \approx \frac{V[3(s + b)^2 + \Omega^2]}{Ts(s + a)(s + b)(s + b + j\Omega)(s + b - j\Omega)} \tag{6.30}$$

Equation (6.30) is rational in s and can be inverted using well-known techniques. The result will be an approximation to $\{y(t)\}$ for all $t \geq 0$. The approximation improves if the system cut-off frequency is lowered or if more terms from the infinite summation are used. The main drawback of the method lies in the difficulty of obtaining a satisfactory error measure on the approximation.

[3] W. K. Linvill, "Sampled-Data Control Systems Studied through Comparison of Sampling with Amplitude Modulation," *Trans. AIEE*, vol. 70, part 2, pp. 1779–1788, 1951.

[4] Cf. Section 4.3.

6.6 DETERMINATION OF SAMPLES OF OUTPUT, $y(kT)$

When a modulated delta-function series of Laplace transform $U^*(s)$ is impressed on a continuous-time system $H(s)$, it is frequently sufficient to determine the output merely at sampling instants instead of continuously, that is, to let $\{y(kT)\}$ serve as an approximation to $\{y(t)\}$. Such a procedure may be desirable since it is usually easier to determine the samples of the output, $\{y(kT)\}$, than the continuous $\{y(t)\}$.

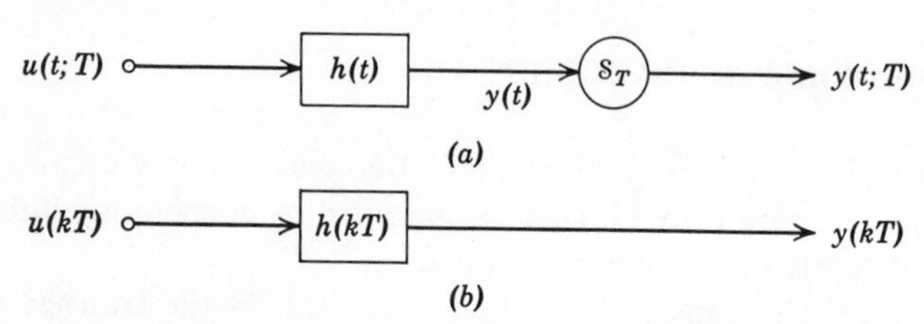

Figure 6-2 Illustration of equivalence between sampled-output continuous-time system (*a*) and discrete-time system (*b*) if input is a discrete-time function.

Referring to Fig. 6-1, the sampled-output transform $Y^*(s)$ is given by

$$Y^*(s) = \frac{1}{T} \sum_{l=-\infty}^{\infty} Y(s + jl\Omega) \tag{6.31}$$

Substituting from (6.27), we have

$$Y^*(s) = \frac{1}{T^2} \sum_{l=-\infty}^{\infty} H(s + jl\Omega) \sum_{p=-\infty}^{\infty} U(s + jp\Omega + jl\Omega) \tag{6.32}$$

However, both l and p range over all integers. Hence

$$Y^*(s) = \frac{1}{T} \sum_{l=-\infty}^{\infty} H(s + jl\Omega) \frac{1}{T} \sum_{p=-\infty}^{\infty} U(s + jp\Omega) \tag{6.33}$$

$$= H^*(s)U^*(s) \tag{6.34}$$

Since both $H^*(s)$ and $U^*(s)$ can be expressed in forms which are rational in e^{sT}, the result of (6.34) is equivalent to that given in (3.14) for a true discrete-time system. We thus observe that the Laplace transform of the *sampled* output of a continuous-time system subjected to a sampled input signal is given by the product of the Laplace transform of the sampled input and the Laplace transform of the sampled weighting function $\{h(t)\}$. Accordingly, we may write

$$Y(z) = H_T(z)U(z) \tag{6.35}$$

where $H_T(z) = \mathcal{Z}[h(kT)]$ is the z transform of the sampled weighting function $\{h(kT)\}$. It follows that for *discrete-time inputs*, a continuous-time system followed by a sampler is indistinguishable from a discrete-time system having a weighting sequence which is equal to the sampled weighting function of the former. This is illustrated in Fig. 6-2.

Example

Suppose we wish to find the output at sampling instants, $\{y(kT)\}$, for a system $H(s) = A/s(s+a)$ subjected to an unspecified discrete-time input $\{u(t; T)\}$. We have

$$Y(z) = H_T(z)U(z)$$

where $U(z) = \mathcal{Z}[u(kT)]$ for $|z| > R_c$ and

$$H_T(z) = \mathcal{Z}\left[\frac{A}{s(s+a)}\right] = \frac{A}{a}\frac{z^{-1}(1-e^{-aT})}{(1-z^{-1})(1-e^{-aT}z^{-1})}, \qquad |z| > e^{-aT}$$

Hence

$$Y(z) = \frac{A}{a}\frac{z^{-1}(1-e^{-aT})}{(1-z^{-1})(1-e^{-aT}z^{-1})}U(z), \qquad |z| > \max(R_c, e^{-aT})$$

and

$$y(kT) = \sum \text{residues of } z^{k-1}Y(z) \text{ at its poles}$$

That is, upon inversion $Y(z)$ will yield the values of the continuous output $y(t)$ at the instants $t = kT$ for $k \geq 0$.

Figure 6-3 Continuous-time system with extrapolated sampled-data input.

The foregoing procedure for determining the sampled output of a continuous-time system is equally applicable if the input sequence is first passed through an extrapolator, as in Fig. 6-3. Thus if we consider a zero-order extrapolator (5.43),

$$Y(z) = \mathcal{Z}\left[\frac{1-e^{-sT}}{s}H(s)\right]U(z) \tag{6.36}$$

$$= (1-z^{-1})\mathcal{Z}\left[\frac{H(s)}{s}\right]U(z) \tag{6.37}$$

Similarly, for a first-order extrapolator (5.49),

$$Y(z) = \mathcal{Z}\left[\frac{(s+1/T)(1-e^{-sT})^2}{s^2}H(s)\right]U(z) \tag{6.38}$$

$$= (1-z^{-1})^2\mathcal{Z}\left[\frac{s+1/T}{s^2}H(s)\right]U(z) \tag{6.39}$$

We have seen that if a modulated delta-function series $\{u(t; T)\}$ is impressed on a continuous-time system having weighting function $\{h(t)\}$, it is a relatively simple matter to find the sampled system output $\{y(kT)\}$. In this derivation the samples of the output $y(kT)$ are obtained at values of t which are of the *same frequency and phase* as those of $\{u(t; T)\}$. The restriction to the same frequency and phase is, however, not necessary, since the output of the system is actually continuous and defined for all time. Thus if we should sample the output at a higher frequency than $2\pi/T$ or at a number of different phase angles within one period T, we would obtain more samples of the output and hence a more precise approximation to the true output $\{y(t)\}$.

6.7 DETERMINATION OF OUTPUT AT INCREASED SAMPLING RATES

Let us consider sampling the output $\{y(t)\}$ in Fig. 6-4 at a frequency $\Omega_2 = 2\pi/T_2$. We have, from (6.27),

$$Y^*(s, T_2) = \frac{1}{T_2}\sum_{l=-\infty}^{\infty} H(s + jl\Omega_2)\frac{1}{T_1}\sum_{p=-\infty}^{\infty} U(s + jp\Omega_1 + jl\Omega_2) \tag{6.40}$$

where $\Omega_1 = 2\pi/T_1$ is the frequency of the samples of $\{u(kT_1)\}$ and $\Omega_2 = 2\pi/T_2$ is the sampling frequency of the output $\{y(kT_2)\}$. If we let $\Omega_1 = \Omega_2$,

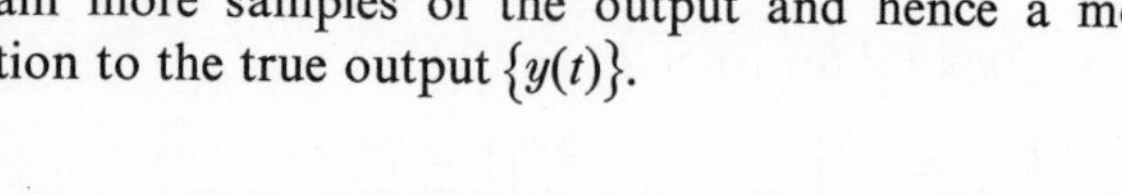

Figure 6-4 Use of different sampling rates for input and output.

we of course obtain (6.33). Now, however, let $\Omega_2 = \nu\Omega_1$, where ν is a positive integer. Then

$$Y^*(s, T_2) = \frac{1}{T_2}\sum_{l=-\infty}^{\infty} H(s + jl\Omega_2)\frac{1}{T_1}\sum_{p=-\infty}^{\infty} U(s + jp\Omega_2/\nu) \tag{6.41}$$

where we have replaced $p\Omega_1 + l\Omega_1\nu$ by simply $p\Omega_1 = p\Omega_2/\nu$, since p ranges over all integers. If we now define $z_1 = e^{sT_1}$ and $z_2 = e^{sT_2}$, such that $z_1 = z_2^{\nu}$, we may write

$$Y(z_2) = H_T(z_2)U(z_1) \tag{6.42}$$

$$= H_T(z_2)U(z_2^{\nu}) \tag{6.43}$$

If there is a zero-order extrapolator at the input of the system we have, from (6.37),

$$Y(z_2) = (1 - z_1^{-1})\mathcal{Z}_2\left[\frac{H(s)}{s}\right]U(z_1) \tag{6.44}$$

$$= (1 - z_2^{-\nu})\mathcal{Z}_2\left[\frac{H(s)}{s}\right]U(z_2^{\nu}) \tag{6.45}$$

Inverting an expression such as (6.43) or (6.45) will give the output values $y(kT_2)$; that is, the interval between output samples will be only $T_2 = T_1/\nu$ instead of T_1. By selecting an arbitrarily large ν, we can have $\{y(t; T_2)\}$ approach $\{y(t)\}$ as closely as desired. In practice, values of ν of 2 or 3 are usually adequate. For larger values of ν, the added effort in inverting $Y(z_2)$ may not be worthwhile.

If $\{h(t)\}$ or $H(s)$ are given, the z transform $\mathcal{Z}_2[H(s)]$ or $\mathcal{Z}_2[H(s)/s]$ can be found in the usual way. Occasionally, however, the z transform with respect to T_2 can be obtained directly from a z transform with respect to T_1. Thus, given $H_{T1}(z_1)$, we expand it in partial fractions to obtain terms of the form

$$\frac{a_i}{1 - b_i z_1^{-1}} = a_i \mathcal{Z}_1[b_i^k] \tag{6.46}$$

$$\frac{a_i b_i z_1^{-1}}{(1 - b_i z_1^{-1})^2} = \frac{a_i}{T_1}\mathcal{Z}_1[kT_1 b_i^k] \tag{6.47}$$

Let $b_i = u_i^{\alpha T_1}$ and $t = k_1 T_1$ where u_i and α are constants. Then

$$\frac{a_i}{1 - b_i z_1^{-1}} = a_i \mathcal{Z}_1[u_i^{\alpha T_1 k_1}] = a_i \mathcal{Z}_1[u_i^{\alpha t}] \tag{6.48}$$

$$\frac{a_i b_i z_1^{-1}}{(1 - b_i z_1^{-1})^2} = \frac{a_i}{T_1}\mathcal{Z}_1[k_1 T_1 u_i^{\alpha T_1 k_1}] = \frac{a_i}{T_1}\mathcal{Z}_1[t u_i^{\alpha t}]. \tag{6.49}$$

Now for $T_2 = T_1/\nu$,

$$a_i \mathcal{Z}_2[u_i^{\alpha t}] = \frac{a_i}{1 - u_i^{\alpha T_2} z_2^{-1}} = \frac{a_i}{1 - b_i^{1/\nu} z_2^{-1}} \tag{6.50}$$

$$\frac{a_i}{T_1}\mathcal{Z}_2[t u_i^{\alpha t}] = \frac{a_i}{T_1}\frac{T_2 u_i^{\alpha T_2} z_2^{-1}}{(1 - u_i^{\alpha T_2} z_2^{-1})^2} = \frac{a_i}{\nu}\frac{b_i^{1/\nu} z_2^{-1}}{(1 - b_i^{1/\nu} z_2^{-1})^2} \tag{6.51}$$

Hence if

$$\mathcal{Z}_1[h_i(t)] = \frac{a_i}{1 - b_i z_1^{-1}}, \quad \text{then } \mathcal{Z}_2[h_i(t)] = \frac{a_i}{1 - b_i^{1/\nu} z_2^{-1}} \tag{6.52}$$

and if

$$\mathcal{Z}_1[h_i(t)] = \frac{a_i b_i z_1^{-1}}{(1 - b_i z_1^{-1})^2}, \quad \text{then } \mathcal{Z}_2[h_i(t)] = \frac{(a_i/\nu) b_i^{1/\nu} z_2^{-1}}{(1 - b_i^{1/\nu} z_2^{-1})^2}. \tag{6.53}$$

Example
Given

$$\mathcal{Z}_1[h(t)] = \frac{3 - 1.62z_1^{-1} + 0.8019z_1^{-2}}{(1 - 0.36z_1^{-1})(1 - 0.81z_1^{-1})^2}$$

where $z_1 = e^{0.4s}$, find $\mathcal{Z}_2[h(t)]$ where $z_2 = e^{0.2s}$. By expanding in partial fractions,

$$\mathcal{Z}_1[h(t)] = \frac{3}{(1 - 0.36z_1^{-1})} + \frac{4(0.81)z_1^{-1}}{(1 - 0.81z_1^{-1})^2}$$

Now $\nu = T_1/T_2 = 2$. Hence from (6.52) and (6.53),

$$\mathcal{Z}_2[h(t)] = \frac{3}{1 - 0.6z_2^{-1}} + \frac{2(0.9)z_2^{-1}}{(1 - 0.9z_2^{-1})^2} = \frac{3 - 3.6z_2^{-1} + 1.35z_2^{-2}}{(1 - 0.6z_2^{-1})(1 - 0.9z_2^{-1})^2}$$

6.8 EXACT INVERSION OF $Y(s)$

We shall now consider an exact method for finding the continuous output of a system subjected to a delta-function series type of input. Let

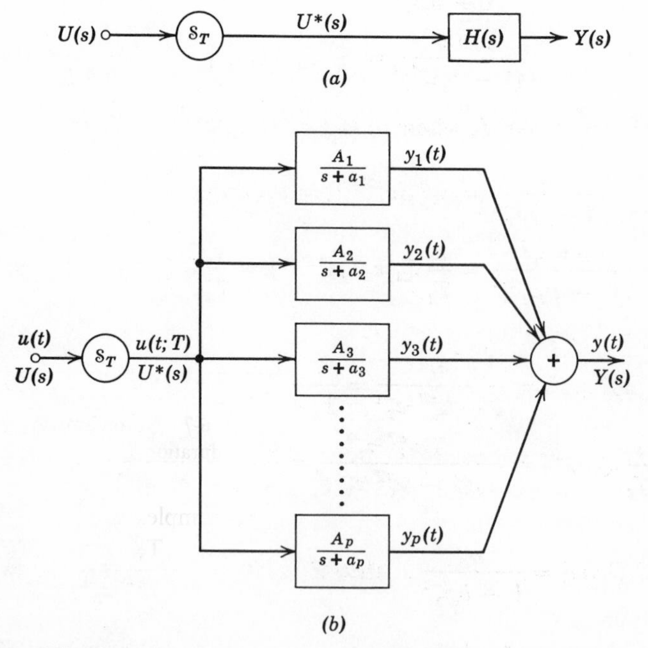

Figure 6-5 Block diagram based on partial-fraction expansion of system transfer function.

the input be given by $U^*(s)$ and the system by $H(s)$, where $H(s)$ is a rational function of s, with the degree of the denominator greater by at least unity than that of the numerator. The transfer function $H(s)$ given in Fig. 6-5*a* can be expanded in partial fractions and then represented by the block diagram of Fig. 6-5*b*. We have, of course,

$$y(t) = \sum_{i=1}^{p} y_i(t) \tag{6.54}$$

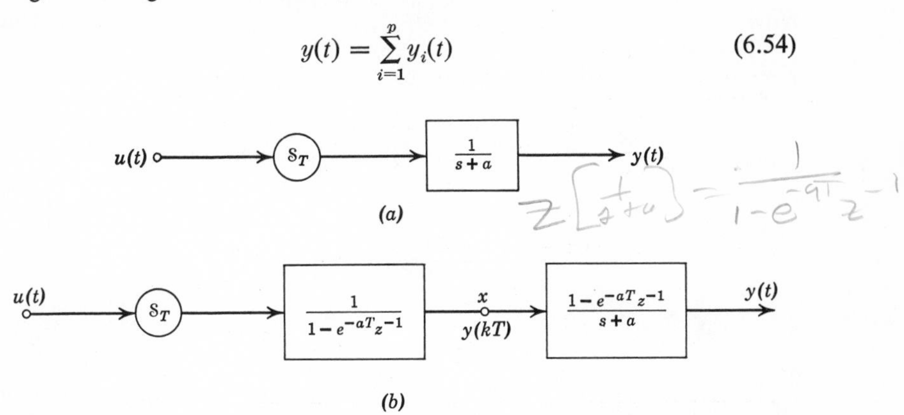

Figure 6-6 Expansion of transfer function into convolution of impulses and normal-mode pulses.

The output of the system of Fig. 6-5*a* will be determined by first determining the outputs of each of the parallel subsystems of Fig. 6-5*b*.

Consider the simple channel shown in Fig. 6-6*a*. The system transfer function can be separated into two cascaded parts, one of which generates a delta-function series (i.e., it is purely discrete) and the other of which represents the transfer function of a pulse-type weighting function. This is illustrated in Fig. 6-6*b*. The pulse-type weighting function is identical to the normal-mode response of the system of Fig. 6-6*a* except that it abruptly goes to zero after T units of time, as shown in Fig. 6-7.

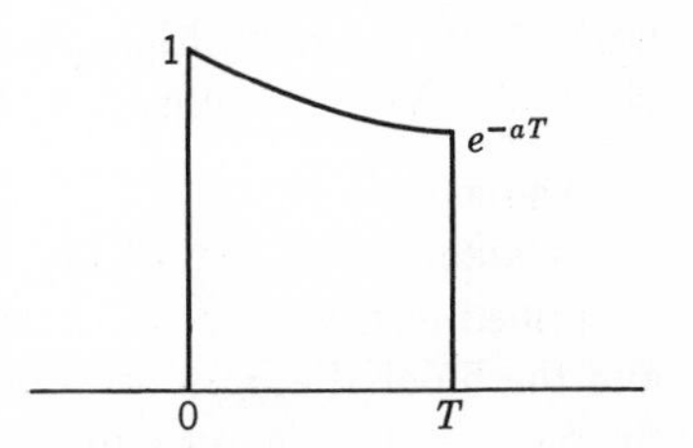

Figure 6-7 Normal-mode pulse e^{-at} of duration T.

It is apparent that the overall transfer functions of Figs. 6-6*a* and 6-6*b* are identical. However, in Fig. 6-6*b* we obtain the samples of the output, $y(kT)$, at the point x, as the reader can readily verify. The normal-mode pulse then "fills in" the interval between samples and provides the continuous output $\{y(t)\}$.

This type of separation into cascade units, the first of which is discrete and the second of which is a normal-mode pulse, can be carried out for all terms of a partial fraction expansion, including those corresponding to multiple poles.[5] We proceed as follows.

Consider

$$\mathscr{L}\left[\frac{At^{\alpha-1}}{(\alpha-1)!}e^{-at}\right] = \frac{A}{(s+a)^{\alpha}} \tag{6.55}$$

Now

$$\frac{At^{\alpha-1}e^{-at}}{(\alpha-1)!} \equiv \frac{Ae^{-akT}}{(\alpha-1)!}(kT+t-kT)^{\alpha-1}e^{-a(t-kT)}$$

$$\equiv \frac{Ae^{-akT}}{(\alpha-1)!}e^{-a(t-kT)}\Big[(kT)^{\alpha-1} + (\alpha-1)(kT)^{\alpha-2}(t-kT) \tag{6.56}$$

$$+\frac{(\alpha-1)(\alpha-2)}{2!}(kT)^{\alpha-3}(t-kT)^2 + \cdots$$

$$+ (t-kT)^{\alpha-1}\Big] \tag{6.57}$$

or

$$\boxed{\mathscr{L}^{-1}\left[\frac{A}{(s+a)^{\alpha}}\right] = A\sum_{i=1}^{\alpha}\frac{e^{-akT}(kT)^{\alpha-i}}{(\alpha-i)!}\,\frac{(t-kT)^{i-1}e^{-a(t-kT)}}{(i-1)!}} \tag{6.58}$$

The first part in the summation (6.58) represents a discrete-time function, while the second represents the normal-mode response of the channel. The expansion (6.58) permits us to separate multiple-pole subsystems into cascade units in the same manner as was used for the simple pole of Fig. 6-6*b*. Note that a multiple pole of order α will be represented by α parallel channels. We shall illustrate the method with the following example.

Example

Consider the system of Fig. 6-8. A unit step-function input $U(s) = 1/s$ is applied to the left of the sampler. A partial-fraction expansion is made and the block diagram of Fig. 6-9 is obtained. We now break up each of the parallel subsystems into cascade units. For the simple-pole subsystems we follow the scheme of Fig. 6-6*b*. For the double-pole subsystem we use (6.58):

$$\mathscr{L}^{-1}\left[\frac{-3}{(s+3)^2}\right] = -3[e^{-3kT}(kT)]e^{-3(t-kT)} - 3[e^{-3kT}](t-kT)e^{-3(t-kT)} \tag{6.59}$$

[5] N. W. Trembath, "Random-Signal Analysis of Linear and Nonlinear Sampled-Data Systems," Report 108, Dynamic Analysis and Control Laboratory, Massachusetts Institute of Technology, Cambridge, Mass., June 1957.

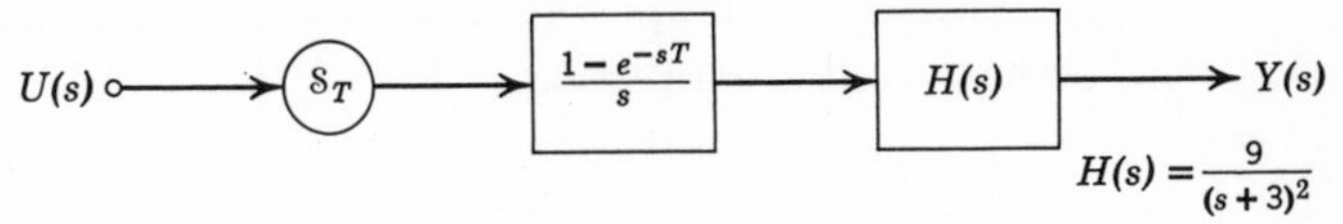

Figure 6-8 Illustrative example for determination of continuous output based on normal-mode response.

Now

$$\mathcal{Z}[e^{-3kT}(kT)] = \frac{Te^{-3T}z^{-1}}{(1 - e^{-3T}z^{-1})^2} \tag{6.60}$$

$$\mathcal{Z}[e^{-3kT}] = \frac{1}{1 - e^{-3T}z^{-1}} \tag{6.61}$$

The transforms of the weighting functions are

$$\mathscr{L}[e^{-3t}] = \frac{1}{s + 3} \tag{6.62}$$

$$\mathscr{L}[te^{-3t}] = \frac{1}{(s + 3)^2} \tag{6.63}$$

The block diagram of Fig. 6-9 can now be redrawn to bring into evidence the purely discrete processing units and the normal-mode networks. This is shown in Fig. 6-10. This diagıam now shows four parallel channels, one each contributed by the two simple-pole channels of Fig. 6-9 and the remaining two contributed by the double-pole channel in accordance with (6.59). The purpose of the function $M(s)$ is to provide a weighting function of finite duration T in the bottom channel, that is, to "shut off" the response to a delta-function input after T seconds.

The samples of the output, $y(kT)$, are given by

$$y(kT) = y_1(kT) + y_2(kT) + y_3(kT) \tag{6.64}$$

The component $y_4(kT)$ does not contribute to the response at $t = kT$

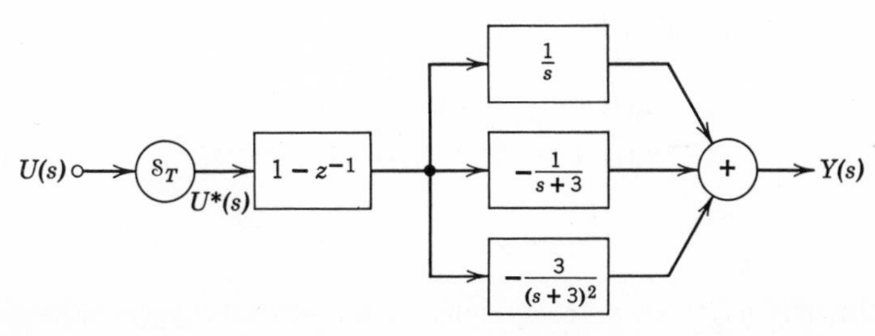

Figure 6-9 Partial-fraction expansion block diagram for system of Fig. 6-8.

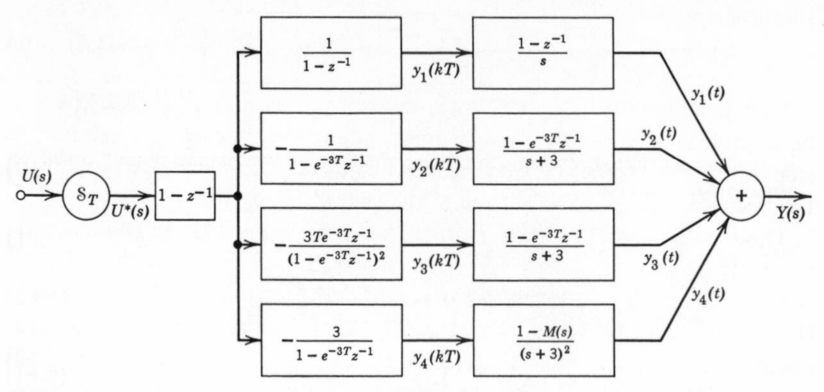

Figure 6-10 Block diagram illustrating separation into cascade elements of discrete processors and normal-mode pulse networks.

because the normal-mode response of the network which is in cascade with it has a weighting function that starts from zero.

We can now write the z transforms $Y_i(z)$, taking into account that $U(z) = 1/(1 - z^{-1})$,

$$Y_1(z) = \frac{1}{1 - z^{-1}}, \qquad y_1(kT) = 1 \tag{6.65}$$

$$Y_2(z) = \frac{-1}{1 - e^{-3T}z^{-1}}, \qquad y_2(kT) = -e^{-3kT} \tag{6.66}$$

$$Y_3(z) = \frac{-3Te^{-3T}z^{-1}}{(1 - e^{-3T}z^{-1})^2}, \qquad y_3(kT) = -3(kT)e^{-3kT} \tag{6.67}$$

$$Y_4(z) = \frac{-3}{1 - e^{-3T}z^{-1}}, \qquad y_4(kT) = -3e^{-3kT} \tag{6.68}$$

Equations (6.65) through (6.68) give the delta-function inputs to the networks whose pulse-type normal-mode responses determine the continuous output. Hence, upon forming the appropriate convolutions,

$$y_1(t) = 1 \tag{6.69}$$

$$y_2(t) = -e^{-3kT}e^{-3(t-kT)} \tag{6.70}$$

$$y_3(t) = -3(kT)e^{-3kT}e^{-3(t-kT)} \tag{6.71}$$

$$y_4(t) = -3e^{-3kT}(t - kT)e^{-3(t-kT)} \tag{6.72}$$

where k is defined so that

$$kT \leq t < kT + T \tag{6.73}$$

The sum of (6.69) through (6.72) is the continuous output $y(t)$ in the interval $kT \leq t < kT + T$. In the particular example, the terms are readily summed and simplified to give

$$y(t) = 1 - e^{-3t} - 3te^{-3t} \quad \text{for all } t \geq 0 \tag{6.74}$$

Such a simple form for the continuous output cannot, of course, always be obtained. The output here is independent of k because we selected a step input to a system that was preceded by a zero-order hold extrapolator. In effect we did not have a discrete-time input at all!

The block diagram of Fig. 6-10 shows a function $M(s)$ in the channel for $y_4(t)$. This function was not specified and did not enter into the calculations. Its sole purpose was to make sure that the response of the "network" $[1 - M(s)]/(s + 3)^2$ be that of a pulse—that is, that the response end abruptly after T seconds. If, for academic reasons, we wish to find $M(s)$, we proceed as follows:

Clearly the combined transfer function for the two lower channels in Fig. 6-10 must be the same as for the bottom channel in Fig. 6-9. Hence

$$-\frac{3}{(s+3)^2} = \frac{-3Te^{-3T}z^{-1}}{(1 - e^{-3T}z^{-1})^2} \times \frac{1 - e^{-3T}z^{-1}}{s+3} - \frac{3[1 - M(s)]}{(1 - e^{-3sT}z^{-1})(s+3)^2} \tag{6.75}$$

Solving for $M(s)$, we obtain

$$M(s) = e^{-3T}z^{-1}(1 + 3T + sT) \tag{6.76}$$

The transfer function in the bottom channel of Fig. 6-10 thus becomes

$$N_4(s) = \frac{1 - [1 + T(s+3)]e^{-3T}z^{-1}}{(s+3)^2} \tag{6.77}$$

$$= \frac{1}{(s+3)^2} - \frac{e^{-3T}e^{-sT}}{(s+3)^2} - \frac{Te^{-3T}e^{-sT}}{s+3} \tag{6.78}$$

Inverting into time domain, we find

$$n_4(t) = te^{-3t}v(t) - e^{-3t}(t - T)v(t - T) - Te^{-3t}v(t - T) \tag{6.79}$$

$$= te^{-3t}[v(t) - v(t - T)] \tag{6.80}$$

where $v(t)$ is the familiar unit step function. Equation (6.80) shows that the weighting function of the network is that of a pulse of duration T, as indeed was expected.

6.9 DETERMINATION OF OUTPUT BY MEANS OF MODIFIED $\mathcal{Z}$ TRANSFORM

In addition to the foregoing method for obtaining the output $\{y(t)\}$ of a continuous-time system to which the input is a modulated delta-function series, we can also use a method which is based on the so-called *modified z transform*.[6] In this method a fictitious time advance δT, $0 \leq \delta < 1$, is introduced as shown in Fig. 6-11. The advanced output is then sampled at $t = kT$, $k = 0, 1, \ldots, \infty$.

The z-transfer function of the system $H(s)e^{\delta Ts}$ is readily obtained from (4.93) by setting l equal to zero,

$$\mathcal{Z}[H(s)e^{\delta Ts}] = \sum \text{residues of } \frac{e^{\delta\lambda T}H(\lambda)}{1 - e^{\lambda T}z^{-1}} \text{ at poles of } H(\lambda) \tag{6.81}$$

The values of the sampled output are then given by

$$y(kT + \delta T) = \mathcal{Z}^{-1}\{U(z)\mathcal{Z}[H(s)e^{\delta Ts}]\} \tag{6.82}$$

for all integers $k \geq 0$. (The symbol $\mathcal{Z}^{-1}$ denotes the inverse z transform operation.)

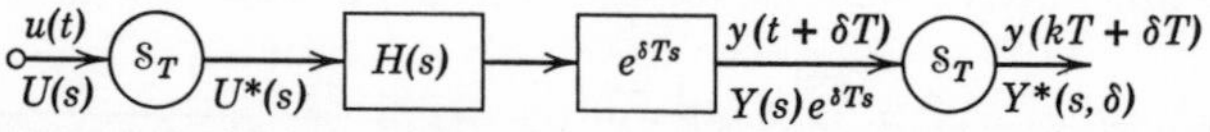

Figure 6-11 Model for modified z-transform determination of $y(t)$.

If δ is permitted to vary continuously from 0 to 1, (6.82) will describe $\{y(t)\}$ for *all* t. For example, if we wish to determine $y(t)$ for $t = 2.7T$, we need merely evaluate (6.82) for $k = 2$ and $\delta = 0.7$.

Note that the modified z transform method can be regarded as a discrete-time solution in which the sampling action is continuously variable in *phase*. In this sense the method is the counterpart of the one described in Section 6.7, in which the sampling *frequency* was taken as variable.

REFERENCES

Jury, E. I., *Sampled-Data Control Systems*, John Wiley and Sons, New York, 1958.

Jury, E. I., *Theory and Application of the Z-Transform Method*, John Wiley and Sons, New York, 1964.

Linvill, William K., "Sampled-Data Control Systems Studied through Comparison of Sampling with Amplitude Modulation," *Trans. AIEE*, vol. 70, part 2, pp. 1779–1788, 1951.

Ragazzini, J. R., and G. Franklin, *Sampled-Data Control Systems*, McGraw-Hill Book Co., New York, 1958.

Trembath, N. W., "Random-Signal Analysis of Linear and Nonlinear Sampled-Data Systems," Report 108, Dynamic Analysis and Control Laboratory, Massachusetts Institute of Technology, Cambridge, Mass., June 1957.

Zadeh, L. A., and C. A. Desoer, *Linear System Theory*, McGraw-Hill Book Co., New York, 1963.

[6] E. I. Jury, "Synthesis and Critical Study of Sampled-Data Control Systems," *Trans. AIEE*, vol. 75, part 2, pp. 141–149, July 1956.

CHAPTER 7

Sampled-Data Control Systems

7.1 INTRODUCTION

Some of the most important applications of discrete-time system theory are found in the field of control. A control system is a system in which feedback is employed to force the output into a precise preselected relation to the input in spite of the action of disturbing forces. To achieve this, both input and output are measured and an error signal is computed which represents the deviation of the output from the currently desired value. Appropriate signals are then supplied to the system's output elements so as to reduce the error. If one or more of the signals are available only at periodically recurring instants of time, the system must be regarded as a sampled-data system. For example, a pulse-type radar system is a sampled-data system because the error signal (that is, the measure of the amount that the radar is off target) is obtainable only at the periodic instants at which the pulse echo is received.

Recent years have seen the increased use of digital computers as components of control systems. Since a digital computer performs discrete numerical calculations, it can accept new input data—as well as supply new output data—only at discrete instants of time. Any system employing such a computer must, therefore, be classed as a sampled-data system. The reasons for which a digital computer may be employed in a control system are that a digital computer is often smaller, lighter, more economical, and may offer greater flexibility than a continuous-time analog computer solving the same control equations. In addition there are some control applications in which the required precision is so great that it can be achieved only with a digital computer.

The presence of signal feedback in a system raises the question of *stability*, and one of the first problems of a system designer is to assure himself that his system will be stable, that is, that it will neither "run away" nor oscillate freely. Next is the problem of transient response, where a system, though stable, may oscillate excessively in response to a step change in input. Alternatively, a system may be too heavily damped and hence have too sluggish a response. The third problem is that of precision and concerns the magnitude of the error between the actual output and the desired output. In general, the greater the required precision, the more difficult the problem of designing a system for a good transient response.

In addition to being concerned with these basic problems of system design, we are also interested in designing systems whose performance is *optimal* with respect to some specified performance criterion. Thus a system may be designed to reduce a step error to zero in *minimum time* or with *minimum total energy consumption*, usually subject to constraints on the available power. Or a system may be designed so that for a given class of "noisy" inputs, the outputs (as functions of time) will differ with minimum mean-square error from some specified desired outputs.[1]

From a theoretical point of view the performance of a sampled-data control system is always inferior to that of a corresponding continuous-time control system. This is so because the sampling action always results in a loss of information. However, sampled-data control systems may occasionally offer practical advantages over continuous-time systems even if they do not contain a digital computer. Thus it is relatively simple to realize pure time delays or time constants of very large value in a sampled-data system.

7.2 BLOCK DIAGRAM REPRESENTATION

System analysts generally find it convenient to represent a system by means of a diagram that depicts a collection of interconnected elementary subsystems. The elementary subsystems are referred to as *blocks* and the diagram which shows the blocks together with their interconnections is called a *block diagram*. A fully labeled block diagram can serve as a complete description of a system. The block diagram representation has much intuitive appeal—often it can provide better first-glance insight into the functional interdependence of the various subsystems than can be obtained from an algebraic description.

Various levels of block diagrams can be drawn, depending on the degree of complexity of the subsystems that are used. The lowest-level block diagram is that in which each integration, time delay, or summation is shown as a distinct subsystem. In higher-level block diagrams, sets of these operations may appear in combined form as single subsystems. A sampled-data system will in addition possess one or more sampling operators to indicate the sampling of continuous-time signals. Extrapolators (usually zero-order hold elements) will be present to provide for the conversion of discrete-time signals to continuous-time signals.

[1] A method for the solution of this problem is given in Chapter 8.

In analyzing a sampled-data system we may employ either the methods based on the use of transfer functions or those based on the use of transition equations. In both cases we frequently find it advantageous first to employ certain block-diagram simplification techniques. These simplification techniques have long been used by system analysts in the study of continuous-time systems. Their extension to discrete-time systems is straightforward. However, some difficulties are encountered when they are applied to sampled-data systems—that is, to systems in which some of the variables are continuous in time and others discrete in time.

If we restrict consideration to sampled-data systems for which all sampling takes place at instants $t = kT$, $k = 0, 1, 2, \ldots, \infty$, we can represent the systems by block diagrams in which only the following types of subsystems are used:

1. Summing points (points at which a linear sum of the inputs is formed).
2. Continuous-time subsystems, denoted by the symbol H_i (where H_i is a transfer function).
3. Discrete-time subsystems, denoted by D_i (where D_i is a transfer function rational in e^{sT}).
4. Sampling operators, denoted by $\mathcal{S}_T$.
5. Hold-type extrapolators, denoted by J_i^r. The superscript r refers to the order of the extrapolator. Thus a zero-order hold is denoted by J_i^0.

In the most primitive (that is, lowest level) type of block diagram the only types of continuous-time and discrete-time subsystems permitted are

$$H_i(s) = \frac{h_i}{s} \tag{7.1}$$

$$D_i(e^{sT}) = d_i e^{-sT} \tag{7.2}$$

In using the transfer function method, we find it advantageous to resolve an extrapolator into two cascaded subsystems, one of which is a discrete-time and the other a continuous-time subsystem.

The uniform sampling operator $\mathcal{S}_T$ converts continuous-time signals $\{f(t)\}$ into discrete-time signals $\{f(kT)\}$. We write

$$\mathcal{S}_T U = U^* \tag{7.3}$$

where the *operand* U is the Laplace transform of a continuous-time function, and U^* is the Laplace transform of the corresponding modulated delta-function series. The latter is, of course, equivalent to the z transform of U if the substitution $z = e^{sT}$ is made.

Some basic block diagram forms involving the uniform sampling operator are shown in Fig. 7-1. The corresponding transfer function relations

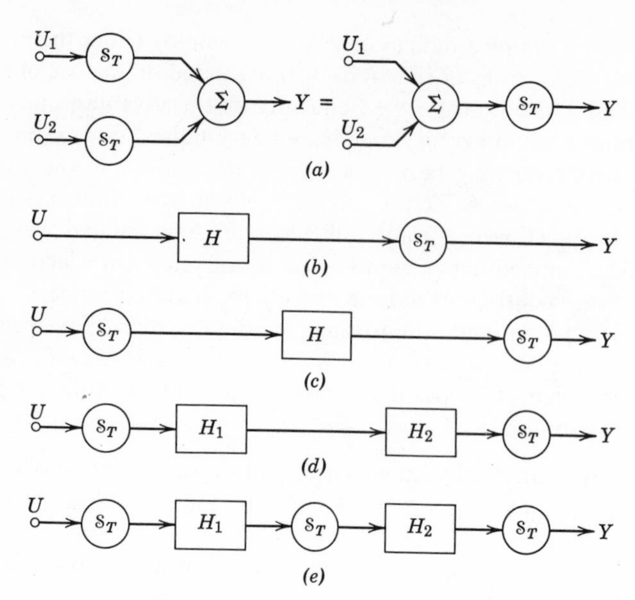

Figure 7-1 Basic block diagram forms.

are as follows:

(*a*) $$Y = \mathcal{S}_T U_1 + \mathcal{S}_T U_2 = \mathcal{S}_T(U_1 + U_2) = U_1{}^* + U_2{}^* \tag{7.4}$$

(*b*) $$Y = \mathcal{S}_T HU = (HU)^* \tag{7.5}$$

(*c*) $$Y = \mathcal{S}_T H \mathcal{S}_T U = H^* U^* \tag{7.6}$$

(*d*) $$Y = \mathcal{S}_T H_2 H_1 \mathcal{S}_T U = (H_2 H_1)^* U^* \tag{7.7}$$

(*e*) $$Y = \mathcal{S}_T H_2 \mathcal{S}_T H_1 \mathcal{S}_T U = H_2{}^* H_1{}^* U^* \tag{7.8}$$

Note that in an algebraic expression, $\mathcal{S}_T$ operates collectively on the components to its *right*.

To utilize the full notational advantage offered by the sampling operator, we broaden the definition of the discrete-time subsystem D to simply an operation that is defined by a difference equation[2]; that is, unless its input is preceded by a sampler, D may accept inputs at any time and not merely at sampling instants. Hence

$$DH = HD \tag{7.9}$$

$$\mathcal{S}_T D = D \mathcal{S}_T \tag{7.10}$$

$$\mathcal{S}_T DH = D \mathcal{S}_T H = DH^* \tag{7.11}$$

[2] An example of such an operation is given by the input–output relation of an electrical network constructed solely with linear amplifiers, summing circuits, and delay lines of delay T. To represent a digital computer, D must be preceded by a sampler.

However,

$$\mathcal{S}_T H = H^* \neq H\mathcal{S}_T. \tag{7.12}$$

We shall illustrate the use of the block diagram simplification technique by determining the output transforms of two types of simple feedback control systems.

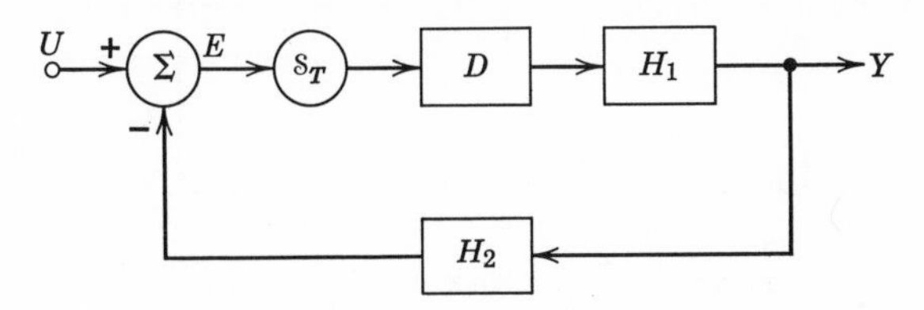

Figure 7-2 Error-sampled feedback system.

Consider the error-sampled feedback system shown in Fig. 7-2. We write

$$Y = H_1 D \mathcal{S}_T E \tag{7.13}$$

$$\mathcal{S}_T E = \mathcal{S}_T(U - H_2 Y) = \mathcal{S}_T U - \mathcal{S}_T H_2 H_1 D \mathcal{S}_T E \tag{7.14}$$

$$E^* = U^* - (H_2 H_1)^* D E^*$$

$$E^* = \frac{U^*}{1 + (H_2 H_1)^* D}$$

$$Y = \frac{H_1 D U^*}{1 + (H_2 H_1)^* D} \tag{7.15}$$

and

$$Y^* = \mathcal{S}_T Y = \frac{H_1^* D U^*}{1 + (H_2 H_1)^* D} \tag{7.16}$$

Next consider the feedback-sampled system illustrated in Fig. 7-3. We have

$$Y = H_2 H_1 E \tag{7.17}$$

$$E = U - H_3 D Y^* \tag{7.18}$$

$$Y^* = \mathcal{S}_T H_2 H_1 (U - H_3 D Y^*) \tag{7.19}$$

$$= (H_2 H_1 U)^* - (H_2 H_1 H_3)^* D Y^*$$

$$= \frac{(H_2 H_1 U)^*}{1 + (H_2 H_1 H_3)^* D} \tag{7.20}$$

$$Y = H_2 H_1 U - \frac{H_2 H_1 H_3 D (H_2 H_1 U)^*}{1 + (H_2 H_1 H_3)^* D} \tag{7.21}$$

It is of interest to note that for the system of Fig. 7-2 it is possible to write transfer functions that relate the input at sampling instants, U^*, to either the continuous-time output, Y, or to the sampled output, Y^*. These transfer functions can be obtained directly from (7.15) and (7.16), respectively. An examination of (7.20) and (7.21), however, shows that no such transfer functions are obtainable for the system of Fig. 7-3.

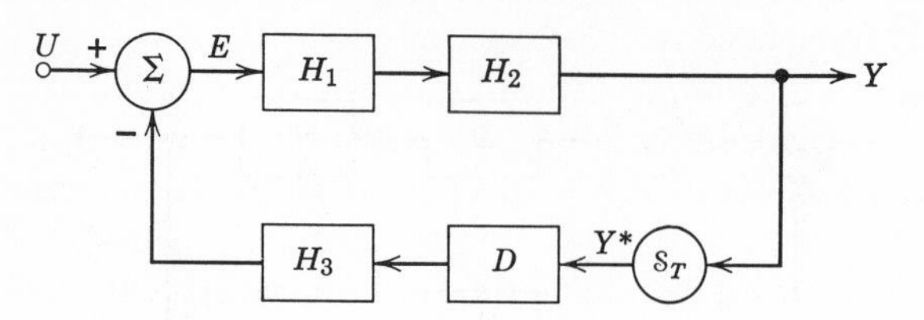

Figure 7-3 Feedback-sampled system.

7.3 SIGNAL FLOW GRAPH ANALYSIS

If a control system contains more than one loop, the step-by-step solution followed in (7.17) to (7.21) becomes too tedious. Instead, we use the technique of signal flow graph reduction suggested by Mason[3]. This technique is particularly rapid and in many cases permits us to write the transform of the output by inspection.

According to Mason, the Laplace transfer function from node i to node j is given by the formula

$$H_{ji} = \frac{\sum_{k} A_{ji,k} \Delta_{ji,k}}{\Delta} \tag{7.22}$$

where $A_{ji,k}$ = transfer function of kth forward path from node i to node j

$$\Delta = 1 - \sum_m L_m + \sum_m \prod_2 L_m - \sum_m \prod_3 L_m + \cdots$$

L_m = transfer function of mth loop.

$\prod_\nu L_m$ = product of ν loop transfer functions that are associated with ν loops which neither touch nor intersect with each other.

$\sum_m \prod_\nu L_m$ = sum of all possible products of ν "nontouching" loop transfer functions.

$\Delta_{ji,k}$ = same as Δ except that only those loops which do not touch or intersect with the kth forward path from node i to node j are considered.

Unfortunately (7.22) applies only to flow graphs which have all branch gains expressed as true transfer functions (Laplace transforms or z transforms). The formula does *not* hold for flow graphs which contain sampling

[3] S. J. Mason, "Feedback Theory—Further Properties of Signal Flow Graphs," *Proc. IRE*, vol. 44, no. 7, pp. 920–926, July 1956.

operators such as the operator $\mathcal{S}_T$. Hence before (7.22) can be applied to the flow graph of a sampled-data system, the graph must be redrawn so as to eliminate all sampling operators. The following method achieves this.[4]

1. Eliminate any obviously redundant sampling operators by inspection. Number the samplers consecutively from 1 to q.
2. Make a new diagram by drawing a node to correspond to each of the q samplers of the given signal flow graph and number the nodes accordingly.
3. Regard all samplers in the given flow graph as representing *open* circuits.
4. For each sampler, determine the Laplace transfer function from the output of the sampler to the input of every other sampler (as well as to the sampler itself). Star (*) each of these transforms since they represent discrete-time subsystems. For each such transfer function, draw a directed branch between the respective nodes in the new flow graph and enter the transfer function alongside this branch.
5. Draw the input nodes and set the inputs at each of these nodes equal to 1. From each input node draw branches to all the sampler nodes to which corresponding paths exist in the given flow path. Assign to these branches the appropriate starred transfer functions $(H_{sr}U_r)^*$, $r = 1, 2, \ldots, m$, $s = 1, 2, \ldots, q$.
6. Add the output nodes and label the outputs Y_j, $j = 1, 2, \ldots, p$. For each path in the given flow graph from a sampler or input node to an output node, draw a corresponding directed branch in the new diagram. Label the branches with the appropriate (nonstarred) Laplace transfer functions.
7. The input nodes can now all be combined into a single unit-input node. Assign the number 0 to this node and then consecutively number the output nodes from $q + 1$ to $q + p$.

The new flow graph is now in a form which permits the use of formula (7.22) to obtain an expression for the output transforms Y_j, $j = 1, 2, \ldots, p$. If the transform of a signal at any other point of the system is desired, it is merely necessary to regard this point as an "output" and to follow the same procedure.

Since the output transforms will be neither rational in s nor in e^{sT}, their inversion into the time domain will require use of the techniques described in Chapter 6. To apply these techniques, *we need merely factor the output transform into two parts such that one can be regarded as a discrete-time input transform and the other as the transfer function of a continuous-time system.*

[4] G. G. Lendaris, "Input-Output Relationships for Multisampled Loop Systems," *Proc. IRE*, vol. 49, no. 11 (correspondence section), November 1961.

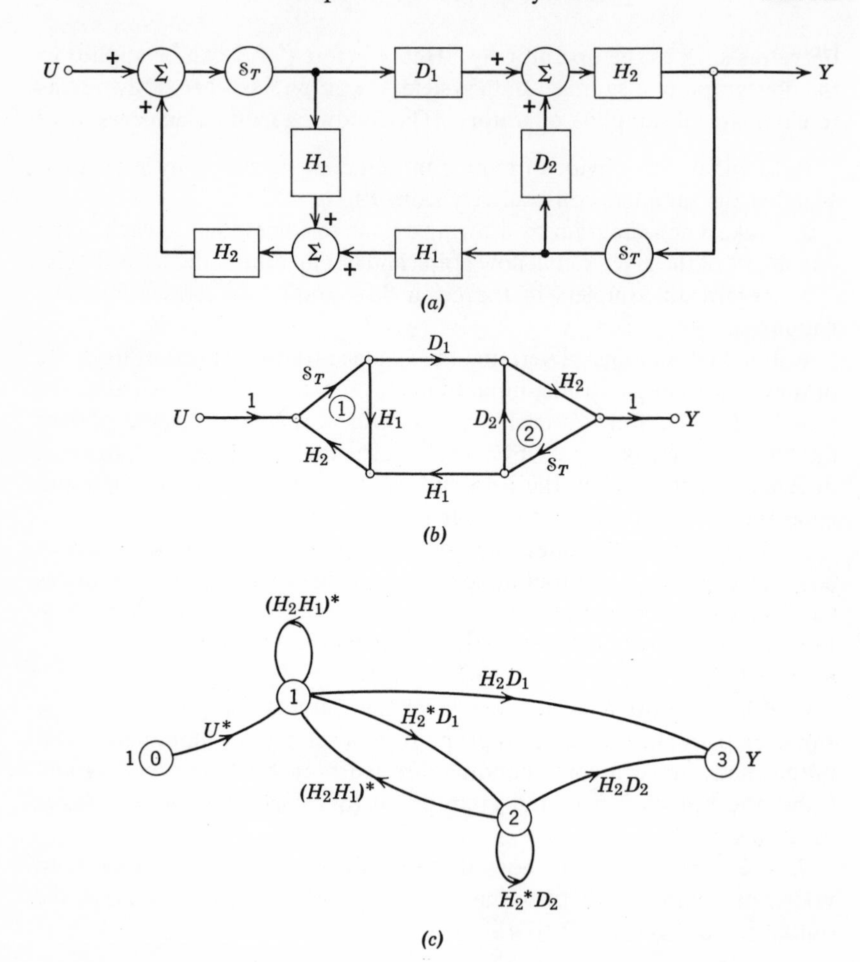

Figure 7-4 Block diagram and flow graphs for Example 1.

Example 1

Consider the block diagram of Fig. 7-4*a*. The corresponding signal flow graph is shown in Fig. 7-4*b*. If we apply the foregoing modification technique to remove the samplers, we obtain the flow graph of Fig. 7-4*c*, from which the forward-path transfer functions are

$$A_{30,1} = H_2D_1U^*$$
$$A_{30,2} = H_2D_2H_2{}^*D_1U^*$$

The loop transfer functions are

$$L_1 = (H_2H_1)^*, \quad L_2 = H_2{}^*D_2, \quad L_3 = H_2{}^*D_1(H_2H_1)^*$$

Hence

$$\Delta = 1 - [(H_2H_1)^* + H_2{}^*D_2 + H_2{}^*D_1(H_2H_1)^*] + (H_2H_1)^*H_2{}^*D_2$$

$$\Delta_{30,1} = 1 - H_2{}^*D_2$$

$$\Delta_{30,2} = 1$$

Then, from (7.22) and after some simplification,

$$Y = \frac{H_2D_1U^*}{1 - (H_2H_1)^* - H_2{}^*D_2 - H_2{}^*D_1(H_2H_1)^* + (H_2H_1)^*H_2{}^*D_2}$$

Example 2

We are to find an expression for the transform of the output Y_2 of the system shown in Fig. 7-5*a*.

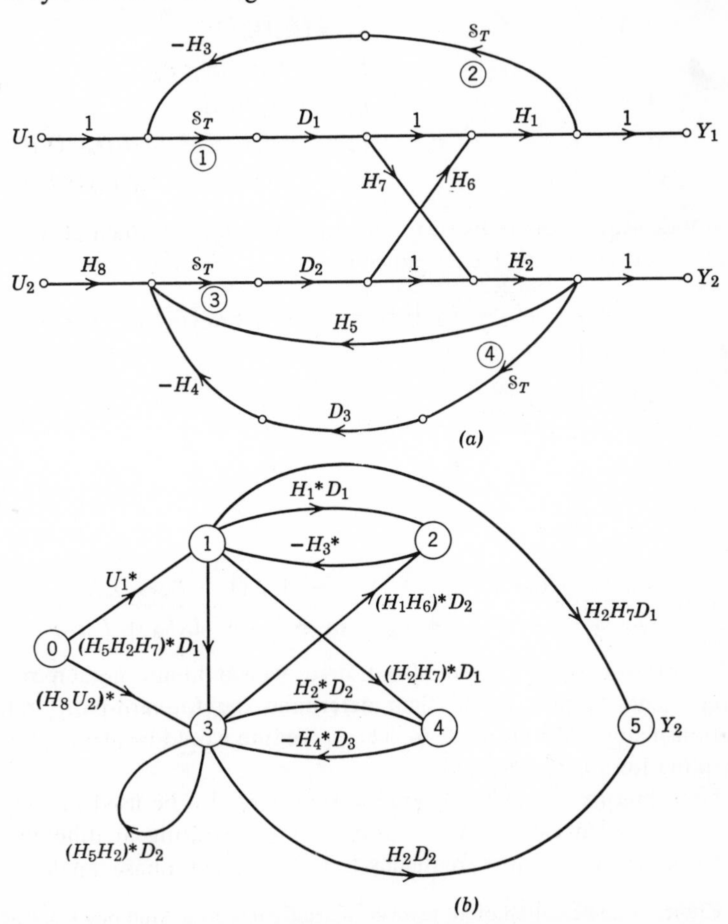

Figure 7-5 System flow graph for Example 2.

We identify the four sampler nodes and draw the flow graph of Fig. 7-5*b*. From Fig. 7-5*b* we can then write

Node Sequences in Forward Paths from Node 0 to Node 5	*Forward-Path Transfer Functions*
0, 3, 5	$A_{50,1} = H_2D_2(H_8U_2)^*$
0, 1, 3, 5	$A_{50,2} = H_2D_2(H_5H_2H_7)^*D_1U_1{}^*$
0, 1, 4, 3, 5	$A_{50,3} = -H_2D_2H_4{}^*D_3(H_2H_7)^*D_1U_1{}^*$
0, 1, 5	$A_{50,4} = H_2H_7D_1U_1{}^*$
0, 3, 2, 1, 5	$A_{50,5} = -H_2H_7D_1H_3{}^*(H_1H_6)^*D_2(H_8U_2)^*$

Loop Node Sequences	*Loop Transfer Functions*
3, 3	$L_1 = (H_5H_2)^*D_2$
3, 4, 3	$L_2 = -H_2{}^*D_2H_4{}^*D_3$
1, 2, 1	$L_3 = -H_1{}^*D_1H_3{}^*$
1, 3, 2, 1	$L_4 = -(H_5H_2H_7)^*D_1(H_1H_6)^*D_2H_3{}^*$
2, 1, 4, 3, 2	$L_5 = H_3{}^*(H_2H_7)^*D_1H_4{}^*D_3(H_1H_6)^*D_2$

(The node sequences are listed to facilitate determining which of the loops and forward paths touch each other.)

$$\Delta = 1 - (L_1 + L_2 + L_3 + L_4 + L_5) + (L_1L_3 + L_2L_3)$$

$$\Delta_{50,1} = 1 - L_3$$

$$\Delta_{50,2} = 1$$

$$\Delta_{50,3} = 1$$

$$\Delta_{50,4} = 1 - (L_1 + L_2)$$

$$\Delta_{50,5} = 1$$

Hence

$$Y_2 = \frac{A_{50,1}(1 - L_3) + A_{50,2} + A_{50,3} + A_{50,4}(1 - L_1 - L_2) + A_{50,5}}{1 - (L_1 + L_2 + L_3 + L_4 + L_5) + (L_1L_3 + L_2L_3)}$$

If the transform of Y_1 were also desired, we would add node number 6 to represent Y_1 and would then determine the forward-path transfer functions from node 0 to node 6. The procedure would be identical to the determination of Y_2.

The techniques described in this section may also be used to simplify systems in which two or more samplers are operating at different frequencies,[5] or at the same frequency but at different phase angles. For

[5] G. Kranc, "Compensation of an Error-Sampled System by a Multirate Controller," *Trans. AIEE*, vol. 76, part 2, pp. 149–159, July 1957.

samplers operating at different phases, we select one sampler to represent "zero" phase and place an appropriate fractional-period delay ahead of each different-phase sampler so as to cancel the particular phase-difference. A corresponding fractional-period advance must then follow each sampler so that the net delay through each of the samplers is not altered. The original, different-phase samplers may now all be replaced by samplers operating at the single, selected phase.

If a system contains two or more samplers operating at different frequencies, we must select a new frequency which is equal to the least multiple of all the sampler frequencies. The sampler-to-sampler node transfer functions may then all be expressed in terms of this one sampling frequency, using the technique of Section 6.7.

7.4 TRANSITION EQUATIONS FOR SAMPLED-DATA SYSTEMS

In Chapter 2 we derived the transition equation for the state variables of a discrete-time system, and in Chapter 6 we derived a similar equation for a continuous-time system.[6] Neither of these equations can be used directly for the description of a sampled-data system. A sampled-data system is a continuous-time system in which some of the variables can change at only discrete instants of time; in addition to continuous-time subsystems, it contains samplers, extrapolators, and (usually) discrete-time subsystems. If a sampled-data system is represented by means of state variables, some of the state variables will vary continuously in time while others will vary discretely. The transition equation for a sampled-data system must, therefore, be a kind of composite of the transition equations of discrete-time and continuous-time systems.

A sampled-data system is usually given in the form of a block diagram in which the various subsystems are described either by transfer functions or by individual state equations. In general it is advantageous to convert the block diagram to a signal flow graph in which each branch represents one of the following operations:

1. amplification, μ_{ij}
2. integration, μ_{ij}/s
3. time delay, $\mu_{ij}e^{-sT}$ (also denoted by $\mu_{ij}E^{-1}$)
4. sampling
5. sampling followed by zero-order hold[7]

[6] Equations (2.29) and (6.14), respectively.

[7] Higher-order hold extrapolations can always be expressed in terms of these five basic operations.

where μ_{ij} is a gain constant (a real number) associated with the directed branch from node j to node i. Such a "primitive" representation is precisely what would result if we were to make an analog computer simulation diagram of a sampled-data system.

To avoid the possibility of omitting any noncontrollable or nonobservable subsystems from the overall system description, we shall base the selection of the state variables directly on the signal flow graph.[8] Specifically, we assign to the system state vector $\mathbf{x}$ as many dimensions as there are "memory cells" in the system. Clearly, one such memory cell will be associated with every integration, with every time-delay element, and with every zero-order hold element. The order of the system then will be given by

$$n = \alpha + \beta + \gamma \tag{7.23}$$

where n = total number of state variables
α = total number of integrators
β = total number of time-delay elements
γ = total number of zero-order hold elements

The n-dimensional state vector of the system can be conveniently partitioned into three component subvectors,

$$\mathbf{x}(t) = \begin{bmatrix} \mathbf{x}^{\alpha}(t) \\ \mathbf{x}^{\beta}(t) \\ \mathbf{x}^{\gamma}(t) \end{bmatrix} \tag{7.24}$$

where $\mathbf{x}^{\alpha}(t)$, $\mathbf{x}^{\beta}(t)$, and $\mathbf{x}^{\gamma}(t)$ are the state vectors that at time t are associated with the continuous-time subsystems, discrete-time subsystems, and zero-order hold elements, respectively. The vector $\mathbf{x}$ will undergo continuous transitions due to the continuous-time subsystems represented by $\mathbf{x}^{\alpha}$, and discontinuous transitions due to the discrete-time subsystems and hold elements represented, respectively, by $\mathbf{x}^{\beta}$ and $\mathbf{x}^{\gamma}$. We shall now examine the nature of these transitions in some detail.

A generalized subvector state-transition diagram for a sampled-data system is shown in Fig. 7-6. The state variables associated with the continuous-time subsystems continuously undergo transitions. At certain discrete instants one or more of these state variables are sampled and the resulting samples are supplied as inputs to either discrete-time subsystems or to zero-order hold elements. The discrete-time subsystems or hold elements will thereupon undergo a discontinuous change in their state. The inputs to the continuous-time subsystems are derived from the inputs to the system as well as from the outputs of the zero-order hold elements.

[8] Cf. Section 2.8.

The discrete-time subsystems obtain their inputs at certain sampling instants from the continuous-time subsystems, from the zero-order holds, or from the system inputs. In addition to undergoing state transitions at these discrete sampling instants, the discrete-time subsystems may, of course, also change state at instants determined by their internal discrete-time "clock."

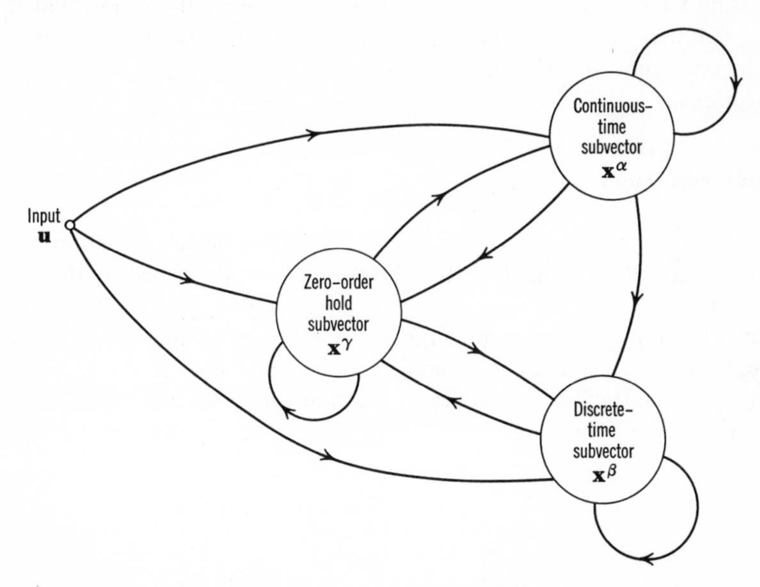

Figure 7-6 Subvector state-transition diagram for sampled-data systems.

At specific sampling instants, the zero-order hold elements sample the inputs as well as the state variables of both the continuous-time and discrete-time subsystems. The resulting samples are stored in the memory cells of the hold elements until they are replaced by new samples. The stored samples—represented by the state variables of the zero-order hold elements—are provided as inputs to the continuous-time and discrete-time subsystems.

In the completely general case, the state variables of the continuous-time and discrete-time subsystems may all be sampled at different instants, or any combination of them may be sampled simultaneously. This is also true of the sampling of the inputs. Further, the sampling of any particular state variable or input may recur either periodically or completely aperiodically. If sampling recurs aperiodically, we have, of course, a nonstationary system, even though all the state equation matrices may be constant matrices. A sampled-data system is stationary only if all its state-equation matrices are constant *and* if the sampling of any particular variable recurs periodically.[9]

We shall now derive three transition equations to describe the three different kinds of transitions that the system state vector $\mathbf{x}$ may undergo. The first of these will refer to transitions that originate with changes in the state variables of the zero-order hold elements, the second will describe the state transitions originating in the discrete-time subsystems, and the third will describe, as a continuous function of time, the state transitions of the continuous-time elements between any two successive discrete-time transitions.[10]

Zero-Order Hold Elements

Let us suppose that at some instant t_k sampling occurs simultaneously at the r zero-order hold elements whose states are described by the state variables x_{q_i}, $i = 1, 2, \ldots, r$, where $\alpha + \beta + 1 \leq q_i \leq n$ for all i. The state variables x_{q_i} will then undergo a discontinuous change at t_k, but all other x_l, $l \neq q_i$, will remain unchanged. The new values of the x_{q_i} will be linear combinations of the values of the state variables and of the inputs at time t_k. We write

$$x_l(t_k^+) = \sum_{j=1}^{n} w_{lj} x_j(t_k) + \sum_{j=1}^{m} b_{lj} u_j(t_k) \quad \text{for } l = q_i \tag{7.25}$$

$$= x_l(t_k) \quad \text{for } l \neq q_i \tag{7.26}$$

$$l = 1, 2, \ldots, n$$

$$\alpha + \beta + 1 \leq q_i \leq n$$

$$i = 1, 2, \ldots, r$$

The w_{lj} and b_{lj} are the constants required to represent the particular state transition under consideration.

In matrix form this can be written as

$$\mathbf{x}(t_k^+) = \mathbf{A}^{\gamma}_{q_1 \cdots q_r} \mathbf{x}(t_k) + \mathbf{B}^{\gamma}_{q_1 \cdots q_r} \mathbf{u}(t_k) \tag{7.27}$$

[9] In a strict sense, *all* sampled-data systems are *nonstationary*. When we refer to a "stationary sampled-data system," we use the term in the sense of "periodically stationary" and limit the nonstationarity to the sampling action.

[10] The approach taken here is based on the work of R. E. Kalman and J. E. Bertram, "A Unified Approach to the Theory of Sampling Systems," *J. Franklin Inst.*, vol. 267, pp. 405–436, May 1959.

where

$$\mathbf{A}^{\gamma}_{q_1\cdots q_r} = \left[\begin{array}{c|c|c} \mathbf{I} & \mathbf{0} & \mathbf{0} \\ \hline \mathbf{0} & \mathbf{I} & \mathbf{0} \\ \hline \mathbf{A}^{\gamma\alpha}_{q_1\cdots q_r} & \mathbf{A}^{\gamma\beta}_{q_1\cdots q_r} & \mathbf{A}^{\gamma\gamma}_{q_1\cdots q_r} \end{array}\right] \tag{7.28}$$

and

$$\mathbf{B}^{\gamma}_{q_1\cdots q_r} = \left[\begin{array}{c} \mathbf{0} \\ \hline \mathbf{0} \\ \hline \mathbf{B}^{\gamma m}_{q_1\cdots q_r} \end{array}\right] \tag{7.29}$$

The elements of $\mathbf{A}^{\gamma\alpha}_{q_1\ldots q_r}$, $\mathbf{A}^{\gamma\beta}_{q_1\ldots q_r}$, $\mathbf{A}^{\gamma\gamma}_{q_1\ldots q_r}$, and $\mathbf{B}^{\gamma m}_{q_1\ldots q_r}$ can be obtained directly from (7.25) and (7.26). *Note that the matrices* $\mathbf{A}^{\gamma}_{q_1\ldots q_r}$ and $\mathbf{B}^{\gamma}_{q_1\ldots q_r}$ *are different for each different set of state variables* x_{q_i}, $\alpha + \beta + 1 \leq q_i \leq n$, $i = 1, 2, \ldots, r$, *that simultaneously undergo a change at a particular instant* t_k.

Example

Consider a system for which $\alpha = 2$, $\beta = 2$, $\gamma = 3$, and $m = 2$. At an instant t_k, the state variable x_6 *alone* undergoes a change. We desire the transition equation that describes this change.

We have

$$n = \alpha + \beta + \gamma = 7$$

$$x_6(t_k^+) = \sum_{j=1}^{6} w_{6j}x_j(t_k) + \sum_{j=1}^{2} b_{6j}u_j(t_k)$$

$$x_l(t_k^+) = x_l(t_k) \quad \text{for all } l \neq 6$$

Hence

$$\mathbf{A}_6^{\gamma} = \left[\begin{array}{cc|cc|ccc} 1 & 0 & 0 & 0 & 0 & 0 & 0 \\ 0 & 1 & 0 & 0 & 0 & 0 & 0 \\ \hline 0 & 0 & 1 & 0 & 0 & 0 & 0 \\ 0 & 0 & 0 & 1 & 0 & 0 & 0 \\ \hline 0 & 0 & 0 & 0 & 1 & 0 & 0 \\ w_{61} & w_{62} & w_{63} & w_{64} & w_{65} & w_{66} & w_{67} \\ 0 & 0 & 0 & 0 & 0 & 0 & 1 \end{array}\right]$$

and

$$\mathbf{B}_6{}^{\gamma} = \begin{bmatrix} 0 & 0 \\ 0 & 0 \\ \hline 0 & 0 \\ 0 & 0 \\ \hline 0 & 0 \\ b_{61} & b_{62} \\ 0 & 0 \end{bmatrix}$$

The desired transition equation is given by

$$\mathbf{x}(t_k{}^+) = \mathbf{A}_6{}^{\gamma}\mathbf{x}(t_k) + \mathbf{B}_6{}^{\gamma}\mathbf{u}(t_k)$$

Discrete-Time Subsystems

We next consider the transitions of the components of $\mathbf{x}^{\beta}$, that is, the transitions of the state variables associated with the discrete-time subsystems. We assume that at some instant t_k a total of s such state variables x_{q_i}, $i = 1, \ldots, s$, undergo a change, where $\alpha + 1 \leq q_i \leq \alpha + \beta$ for all i. All other state variables x_l, $l \neq q_i$, remain unchanged. We write

$$x_l(t_k{}^+) = \sum_{j=1}^{n} w_{lj}x_j(t_k) + \sum_{j=1}^{m} b_{lj}u_j(t_k) \quad \text{for } l = q_i \tag{7.30}$$

$$= x_l(t_k) \quad \text{for } l \neq q_i \tag{7.31}$$

$$l = 1, \ldots, n$$

$$\alpha + 1 \leq q_i \leq \alpha + \beta$$

$$i = 1, \ldots, s$$

In matrix form

$$\mathbf{x}(t_k{}^+) = \mathbf{A}^{\beta}_{q_1 \cdots q_s}\mathbf{x}(t_k) + \mathbf{B}^{\beta}_{q_1 \cdots q_s}\mathbf{u}(t_k) \tag{7.32}$$

where

$$\mathbf{A}^{\beta}_{q_1 \cdots q_s} = \left[\begin{array}{c|c|c} \mathbf{I} & \mathbf{0} & \mathbf{0} \\ \hline \mathbf{A}^{\alpha\beta}_{q_1 \cdots q_s} & \mathbf{A}^{\beta\beta}_{q_1 \cdots q_s} & \mathbf{A}^{\gamma\beta}_{q_1 \cdots q_s} \\ \hline \mathbf{0} & \mathbf{0} & \mathbf{I} \end{array}\right] \tag{7.33}$$

and

$$\mathbf{B}^{\beta}_{q_1 \cdots q_s} = \begin{bmatrix} \mathbf{0} \\ \hline \mathbf{B}^{\beta m}_{q_1 \cdots q_s} \\ \hline \mathbf{0} \end{bmatrix} \tag{7.34}$$

The elements of the submatrices in (7.33) and (7.34) can be obtained directly from (7.30).

Example

Consider a system for which $\alpha = 2$, $\beta = 2$, $\gamma = 3$, and $m = 2$. At t_k *both* x_3 and x_4 undergo a change. We write

$$n = \alpha + \beta + \gamma = 7$$

$$\mathbf{x}(t_k^+) = \mathbf{A}^{\beta}_{3,4}\mathbf{x}(t_k) + \mathbf{B}^{\beta}_{3,4}\mathbf{u}(t_k)$$

where

$$\mathbf{A}^{\beta}_{3,4} = \left[\begin{array}{cc|cc|ccc} 1 & 0 & 0 & 0 & 0 & 0 & 0 \\ 0 & 1 & 0 & 0 & 0 & 0 & 0 \\ \hline w_{31} & w_{32} & w_{33} & w_{34} & w_{35} & w_{36} & w_{37} \\ w_{41} & w_{42} & w_{43} & w_{44} & w_{45} & w_{46} & w_{47} \\ \hline 0 & 0 & 0 & 0 & 1 & 0 & 0 \\ 0 & 0 & 0 & 0 & 0 & 1 & 0 \\ 0 & 0 & 0 & 0 & 0 & 0 & 1 \end{array}\right]$$

and

$$\mathbf{B}^{\beta}_{3,4} = \left[\begin{array}{cc} 0 & 0 \\ 0 & 0 \\ \hline b_{31} & b_{32} \\ b_{41} & b_{42} \\ \hline 0 & 0 \\ 0 & 0 \\ 0 & 0 \end{array}\right]$$

If only x_3 were to change at t_k, then instead of $\mathbf{A}^{\beta}_{3,4}$ and $\mathbf{B}^{\beta}_{3,4}$, we would use $\mathbf{A}_3{}^{\beta}$ and $\mathbf{B}_3{}^{\beta}$. The latter would differ from the former in that $w_{4i} = 0$ for all i except $i = 4$, $w_{44} = 1$, and $b_{4j} = 0$ for all j.

Continuous-Time Subsystems

We shall now derive a transition equation for the continuous-time elements of a sampled-data system. Assume that a discontinuous state transition has occurred at time t_k, and that the next such transition will not occur till time t_{k+1}. Then during the interval $(t_k, t_{k+1}]$ the state subvectors $\mathbf{x}^\beta$ and $\mathbf{x}^\gamma$ will remain constant. The subvector $\mathbf{x}^\alpha$, however, will undergo *continuous* transitions. In effect, we have a continuous-time subsystem of dimension α, with state equation

$$\frac{d}{dt}\mathbf{x}^\alpha(t) = \mathbf{A}\mathbf{x}^\alpha(t) + \mathbf{B}\mathbf{u}^\alpha(t) \tag{7.35}$$

Its state vector $\mathbf{x}^\alpha$ during the interval $(t_k, t_{k+1}]$ is a continuous function of the initial state $\mathbf{x}^\alpha(t_k)$ and a functional of the input $\mathbf{u}^\alpha$ to this continuous-time subsystem in the interval $[t_k, t)$. As can be seen from Fig. 7-6, the input to this subsystem consists of a linear combination of the constant state variables $x_l(t_k)$, $l = \alpha + \beta + 1, \ldots, n$ (from the zero-order hold elements), and the components of the system input vector $\mathbf{u}$. We write, using (6.14) and (6.22),

$$\mathbf{x}^\alpha(t_k + v) = \mathbf{\Phi}^c(v)\mathbf{x}^\alpha(t_k) + \mathbf{\Lambda}^{\alpha\gamma}(v)\mathbf{x}^\gamma(t_k) + \int_0^{v-} \mathbf{\Phi}^c(v - \tau)\mathbf{B}^{\alpha m}\mathbf{u}(t_k + \tau)\, d\tau \tag{7.36}$$

where $\mathbf{\Phi}^c$, $\mathbf{\Lambda}^{\alpha\gamma}$, and $\mathbf{B}^{\alpha m}$ are $\alpha \times \alpha$, $\alpha \times \gamma$, and $\alpha \times m$ matrices, respectively.

In terms of the *system* state vector $\mathbf{x}$,

$$\mathbf{x}(t_k + v) = \mathbf{\Psi}(v)\mathbf{x}(t_k) + \int_0^{v-} \mathbf{\Psi}(v - \tau)\mathbf{B}^\alpha\mathbf{u}(t_k + \tau)\, d\tau \tag{7.37}$$

$$0 < v \leq t_{k+1} - t_k \quad \text{for all } k$$

where

$$\mathbf{\Psi}(v) = \left[\begin{array}{c:c:c} \mathbf{\Phi}^c(v) & \mathbf{0} & \mathbf{\Lambda}^{\alpha\gamma}(v) \\ \hdashline \mathbf{0} & \mathbf{I} & \mathbf{0} \\ \hdashline \mathbf{0} & \mathbf{0} & \mathbf{I} \end{array}\right] \tag{7.38}$$

and

$$\mathbf{B}^\alpha = \left[\begin{array}{c} \mathbf{B}^{\alpha m} \\ \hdashline \mathbf{0} \\ \hdashline \mathbf{0} \end{array}\right] \tag{7.39}$$

The matrices $\mathbf{\Psi}(v)$ and $\mathbf{B}^\alpha$ are $n \times n$ and $n \times m$ matrices, respectively. Their elements can be determined in accordance with the procedure of

Section 6.2 from the state equation (7.35) and the *interconnection matrices* $\mathbf{Q}^{\gamma}$ and $\mathbf{Q}^{u}$ which, respectively, relate the components of the subvector $\mathbf{x}^{\gamma}$ and the components of the input $\mathbf{u}$ to the subvector $\mathbf{x}^{\alpha}$; that is,

$$\mathbf{\Phi}^{c}(v) = e^{\mathbf{A}v} \tag{7.40}$$

$$\mathbf{\Lambda}^{\alpha\gamma}(v) = \mathbf{\Lambda}^{c}(v)\mathbf{Q}^{\gamma} = \int_{0}^{v-} \mathbf{\Phi}^{c}(v-\tau)\mathbf{B}\mathbf{Q}^{\gamma}\, d\tau \tag{7.41}$$

$$\mathbf{B}^{\alpha m} = \mathbf{B}\mathbf{Q}^{u} \tag{7.42}$$

The transition equations (7.37), (7.27), and (7.32) enable us to determine the state of any linear sampled-data system with stationary state transition matrices at any time t, given the state at some time $t_0 < t$ and the input over the interval $[t_0, t)$. The equations apply for *any* kind of sampling, from the case where all samplers operate simultaneously with a fixed period, to the case where all samplers operate individually and aperiodically.

7.5 ILLUSTRATIVE EXAMPLE

We shall now illustrate the transition-equation method of analyzing sampled-data systems by applying it to a system in which sampling occurs at nonperiodic instants. Consider the system shown in Fig. 7-7. The continuous-time subsystem H is described by the transfer function

$$H(s) = \frac{K_3}{s(s+4)}$$

The discrete-time subsystem D is described by the difference equation

$$g(k) + 3g(k-1) + 2g(k-2) = 5f(k-1)$$

where f and g are, respectively, the input and output of the discrete-time subsystem. S^1, S^2, and S^3 denote three aperiodic sampling operators. There are two zero-order holds, denoted by $J_1{}^0$ and $J_2{}^0$, and two linear amplifiers with gains K_1 and K_2.

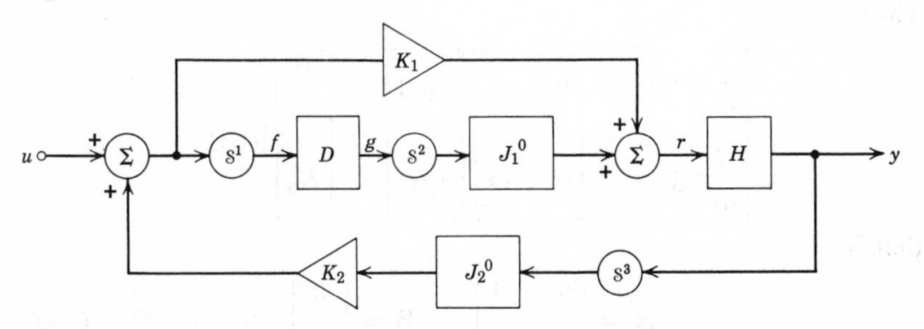

Figure 7-7 Sampled-data control system example.

The following sampling operations occur in the system: let $t_0 < t_1 < t_2 < t_3 < t_4$.

Time	*Operation*
t_1	$\mathcal{S}^3$ inserts new sampled value into J_2^0
t_2	$\mathcal{S}^1$ reads a new input f into D and all internal variables of D change
t_3	$\mathcal{S}^2$ reads output of D into J_1^0
t_4	$\mathcal{S}^1$ reads new input into D

We are given the state of the system at time t_0 and wish to determine the state at time t, where $t_3 < t \leq t_4$. No discontinuous state transitions occur in the interval $[t_0, t)$ other than at t_1, t_2, and t_3.

To solve this problem we must first determine the structure and interconnections of the system. From inspection we note that $\alpha = 2$, $\beta = 2$, $\gamma = 2$, $m = 1$, and $p = 1$. Hence the *order* of the system is $n = 6$.

Let x_1 and x_2 be associated with the continuous-time subsystem H, x_3, and x_4 be associated with the discrete-time subsystem D, and x_5 and x_6 be, respectively, associated with the zero-order holds J_1^0 and J_2^0.

We now determine the state equations for the continuous- and discrete-time systems. We write

$$H(s) = \frac{\mathscr{L}[y(t)]}{\mathscr{L}[r(t)]} = \frac{K_3}{s(s+4)} \tag{7.43}$$

Hence

$$\frac{d^2}{dt^2}\, y(t) + 4\frac{d}{dt}\, y(t) = K_3 r(t) \tag{7.44}$$

Let

$$y(t) = x_1(t) \tag{7.45}$$

$$\frac{dy(t)}{dt} = x_2(t) \tag{7.46}$$

Then

$$\begin{bmatrix} \dfrac{d}{dt}\, x_1(t) \\ \dfrac{d}{dt}\, x_2(t) \end{bmatrix} = \begin{bmatrix} 0 & 1 \\ 0 & -4 \end{bmatrix} \begin{bmatrix} x_1(t) \\ x_2(t) \end{bmatrix} + \begin{bmatrix} 0 \\ K_3 \end{bmatrix} [r(t)] \tag{7.47}$$

that is,

$$\mathbf{A} = \begin{bmatrix} 0 & 1 \\ 0 & -4 \end{bmatrix} \qquad \mathbf{B} = \begin{bmatrix} 0 \\ K_3 \end{bmatrix} \tag{7.48}$$

For the discrete-time subsystem we obtain, using the procedure of (2.65) to (2.71),

$$\begin{bmatrix} x_3(k+1) \\ x_4(k+1) \end{bmatrix} = \begin{bmatrix} -3 & 1 \\ -2 & 0 \end{bmatrix} \begin{bmatrix} x_3(k) \\ x_4(k) \end{bmatrix} + \begin{bmatrix} 5 \\ 0 \end{bmatrix} [f(k)] \tag{7.49}$$

$$g(k) = x_3(k) \tag{7.50}$$

We can utilize (7.45), (7.47), (7.49), and (7.50) to redraw Fig. 7-7 in the form of a detailed signal flow graph, as shown in Fig. 7-8.

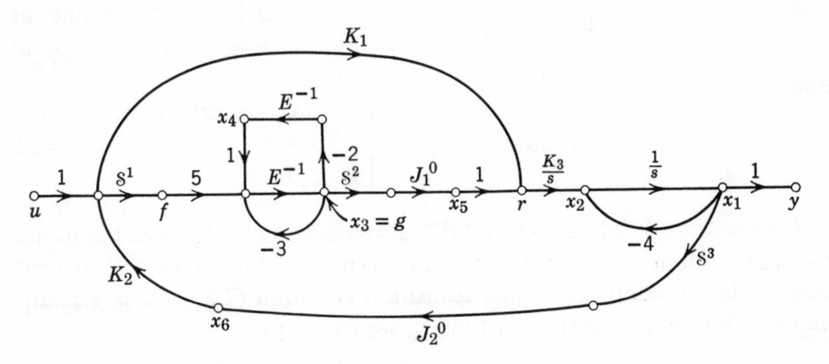

Figure 7-8 Signal flow graph for example.

To determine the transition equation of the continuous-time subsystem we proceed as follows. From (7.48) we can write

$$[s\mathbf{I} - \mathbf{A}]^{-1} = \begin{bmatrix} \frac{1}{s} & \frac{1}{s(s+4)} \\ 0 & \frac{1}{s+4} \end{bmatrix} \tag{7.51}$$

From (6.4) through (6.7),

$$\mathbf{\Phi}^c(v) = \begin{bmatrix} 1 & \frac{1}{4}(1 - e^{-4v}) \\ 0 & e^{-4v} \end{bmatrix} \tag{7.52}$$

From (6.20),

$$\mathbf{\Lambda}^c(v) = \begin{bmatrix} \frac{K_3}{4}(v - \frac{1}{4} + \frac{1}{4}e^{-4v}) \\ \frac{K_3}{4}(1 - e^{-4v}) \end{bmatrix} \tag{7.53}$$

We note from Fig. 7-8 that

$$r = x_5 + K_1 K_2 x_6 + K_1 u \tag{7.54}$$

Hence the applicable interconnection matrices are

$$\mathbf{Q}^{\gamma} = [1 \quad K_1K_2] \tag{7.55}$$

$$\mathbf{Q}^{u} = [K_1] \tag{7.56}$$

We obtain

$$\mathbf{\Lambda}^{\alpha\gamma}(v) = \mathbf{\Lambda}^{c}(v)\mathbf{Q}^{\gamma} = \begin{bmatrix} \frac{K_3}{4}(v - \frac{1}{4} + \frac{1}{4}e^{-4v}) & \frac{K_1K_2K_3}{4}(v - \frac{1}{4} + \frac{1}{4}e^{-4v}) \\ \frac{K_3}{4}(1 - e^{-4v}) & \frac{K_1K_2K_3}{4}(1 - e^{-4v}) \end{bmatrix} \tag{7.57}$$

and

$$B^{\alpha m} = \mathbf{B}\mathbf{Q}^{u} = \begin{bmatrix} 0 \\ K_1K_3 \end{bmatrix} \tag{7.58}$$

The matrices $\mathbf{\Phi}^{c}(v)$, $\mathbf{\Lambda}^{\alpha\gamma}(v)$, and $\mathbf{B}^{\alpha m}$ are sufficient to determine the transition equation matrices $\mathbf{\Psi}(v)$ and $\mathbf{B}^{\alpha}$ as given by (7.38) and (7.39). We can now write the continuous-time transition equation (7.37) for *any* t_k and *any* v. Thus given the state at time t_0, we have at t_1

$$\mathbf{x}(t_1) = \mathbf{\Psi}(t_1 - t_0)\mathbf{x}(t_0) + \int_0^{t_1^- - t_0} \mathbf{\Psi}(t_1 - t_0 - \tau)\mathbf{B}^{\alpha}\mathbf{u}(t_0 + \tau)\, d\tau \tag{7.59}$$

At t_1 there is a discontinuous transition to a new state $\mathbf{x}(t_1^+)$. Then

$$\mathbf{x}(t_2) = \mathbf{\Psi}(t_2 - t_1)\mathbf{x}(t_1^+) + \int_0^{t_2^- - t_1} \mathbf{\Psi}(t_2 - t_1 - \tau)\mathbf{B}^{\alpha}\mathbf{u}(t_1 + \tau)\, d\tau \tag{7.60}$$

Another discontinuous transition now takes us to $x(t_2^+)$, after which

$$\mathbf{x}(t_3) = \mathbf{\Psi}(t_3 - t_2)\mathbf{x}(t_2^+) + \int_0^{t_3^- - t_2} \mathbf{\Psi}(t_3 - t_2 - \tau)\mathbf{B}^{\alpha}\mathbf{u}(t_2 + \tau)\, d\tau \tag{7.61}$$

Following the discontinuous transition at t_3, we have for all $t_3 < t \leq t_4$

$$\mathbf{x}(t) = \mathbf{\Psi}(t - t_3)\mathbf{x}(t_3^+) + \int_0^{t^- - t_3} \mathbf{\Psi}(t - t_3 - \tau)\mathbf{B}^{\alpha}\mathbf{u}(t_3 + \tau)\, d\tau \tag{7.62}$$

The last equation is the desired result. However, before we can evaluate (7.62), we must determine the discontinuous transitions at t_1, t_2, and t_3.

At t_1, J_2^0 obtains a new value of y. This causes the state variable x_6 to undergo a discontinuous transition. All other variables remain unchanged. Since we have let $y = x_1$, we have

$$\mathbf{x}(t_1^+) = \mathbf{A}_6{}^{\gamma}\mathbf{x}(t_1) \tag{7.63}$$

where

$$\mathbf{A}_6^\gamma = \left[\begin{array}{cc:cc:cc} 1 & 0 & 0 & 0 & 0 & 0 \\ 0 & 1 & 0 & 0 & 0 & 0 \\ \hdashline 0 & 0 & 1 & 0 & 0 & 0 \\ 0 & 0 & 0 & 1 & 0 & 0 \\ \hdashline 0 & 0 & 0 & 0 & 1 & 0 \\ 1 & 0 & 0 & 0 & 0 & 0 \end{array}\right] \tag{7.64}$$

Similarly, at t_3 a new value of g is read into J_1^0, causing x_5 to undergo a discontinuous transition. Since we have let $g = x_3$,

$$\mathbf{x}(t_3^+) = \mathbf{A}_5^\gamma \mathbf{x}(t_3) \tag{7.65}$$

where

$$\mathbf{A}_5^\gamma = \left[\begin{array}{cc:cc:cc} 1 & 0 & 0 & 0 & 0 & 0 \\ 0 & 1 & 0 & 0 & 0 & 0 \\ \hdashline 0 & 0 & 1 & 0 & 0 & 0 \\ 0 & 0 & 0 & 1 & 0 & 0 \\ \hdashline 0 & 0 & 1 & 0 & 0 & 0 \\ 0 & 0 & 0 & 0 & 0 & 1 \end{array}\right] \tag{7.66}$$

The discontinuous transition at t_2 is due to simultaneous changes in the state variables x_3 and x_4 of the discrete-time subsystem. From Fig. 7-8,

$$f(t_k) = u(t_k) + K_2 x_6(t_k) \tag{7.67}$$

The elements of the $\mathbf{A}^\beta$ and $\mathbf{B}^\beta$ matrices can now be obtained. We write

$$\mathbf{x}(t_2^+) = \mathbf{A}_{3,4}^\beta \mathbf{x}(t_2) + \mathbf{B}_{3,4}^\beta \mathbf{u}(t_2) \tag{7.68}$$

where, from (7.49) and (7.67),

$$\mathbf{A}_{3,4}^\beta = \left[\begin{array}{cc:cc:cc} 1 & 0 & 0 & 0 & 0 & 0 \\ 0 & 1 & 0 & 0 & 0 & 0 \\ \hdashline 0 & 0 & -3 & 1 & 0 & 5K_2 \\ 0 & 0 & -2 & 0 & 0 & 0 \\ \hdashline 0 & 0 & 0 & 0 & 1 & 0 \\ 0 & 0 & 0 & 0 & 0 & 1 \end{array}\right] \tag{7.69}$$

and

$$\mathbf{B}^{\beta}_{3,4} \begin{bmatrix} 0 \\ 0 \\ \hdashline 5 \\ 0 \\ \hdashline 0 \\ 0 \end{bmatrix} \tag{7.70}$$

Thus given $\mathbf{x}(t_0)$, we use in succession (7.59) to find $\mathbf{x}(t_1)$, (7.63) to find $\mathbf{x}(t_1{}^+)$, (7.60) to obtain $\mathbf{x}(t_2)$, (7.68) to find $\mathbf{x}(t_2{}^+)$, (7.61) to obtain $\mathbf{x}(t_3)$, and (7.65) to find $\mathbf{x}(t_3{}^+)$. Then finally (7.62) yields the state $\mathbf{x}(t)$ for any t in the interval $(t_3, t_4]$.

It is possible to combine all these equations into a *single* transition equation that relates the state at time t, $t_3 < t \leq t_4$, to the state at t_0. The equation unfortunately would be a very complicated difference-integral equation. However, if we limit ourselves to the case where the input is zero during the interval $[t_0, t]$, we can write

$$\begin{aligned} \mathbf{x}(t) &= \boldsymbol{\Psi}(t - t_3)\mathbf{A}_5{}^{\gamma}\boldsymbol{\Psi}(t_3 - t_2)\mathbf{A}^{\beta}_{3,4}\boldsymbol{\Psi}(t_2 - t_1)A_6{}^{\gamma}\boldsymbol{\Psi}(t_1 - t_0)\mathbf{x}(t_0) && (7.71) \\ &= \boldsymbol{\chi}(t, t_0)\mathbf{x}(t_0) && (7.72) \end{aligned}$$

The matrix $\boldsymbol{\chi}(t, t_0)$ is nonstationary since it is dependent on the sampling instants.

7.6 SPECIAL CASES OF SAMPLING AND "HOLD" EXTRAPOLATION

In the foregoing two sections we described a procedure for determining the state of a sampled-data system in which sampling occurred in an aperiodic and nonsynchronized manner. This is the most general case, and from it we can derive the transition equations for a number of special cases, such as, for example, systems in which some or all the samplers operate in synchronism or with fixed periods.

Let us consider the case shown in Fig. 7-9*a* where all samplers in a system operate in synchronism and with a single, uniform period T. At instants kT, $k = 0, 1, \ldots, \infty$, all the state variables associated with the hold elements and the discrete-time subsystems simultaneously undergo a discontinuous transition, thereby changing the system state from $\mathbf{x}(kT)$ to $\mathbf{x}(kT^+)$. In any interval $kT < t \leq kT + T$, the state changes in accordance with a continuous-time transition equation of the form (7.37), where now, however, $t_k = kT$. Thus only two transition equations are required to determine the state at any time t—one for the discontinuous transition from kT to kT^+ and the other for the continuous transitions from kT^+ to $kT + T$.

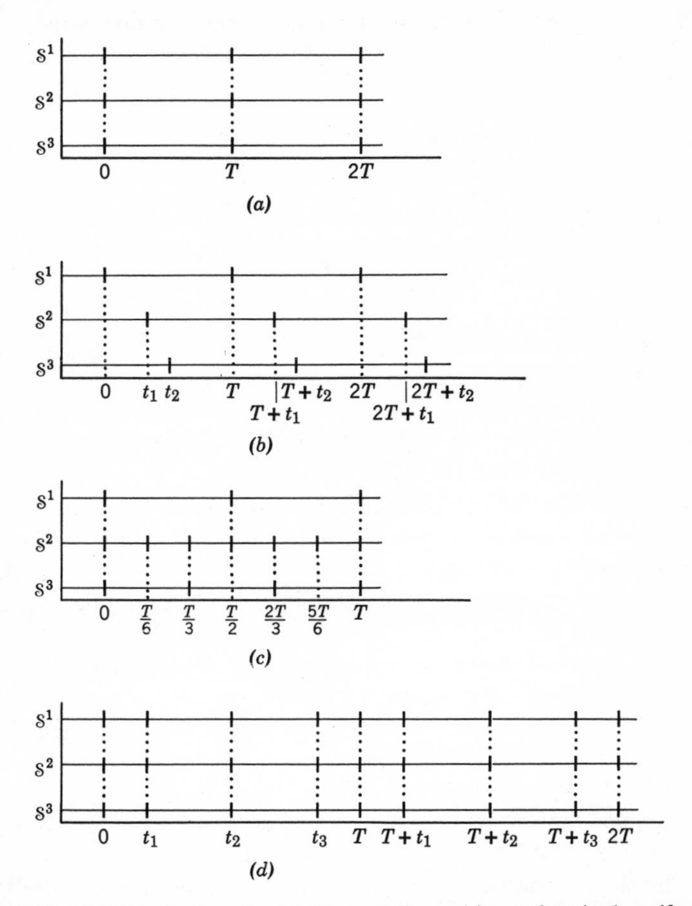

Figure 7-9 Various forms of periodic sampling: (*a*) synchronized, uniform; (*b*) nonsynchronized, uniform; (*c*) multiple rate; (*d*) synchronized, periodic-nonuniform.

In Fig. 7-9*b* all the samplers operate at a uniform period but are not synchronized. We shall need a separate transition equation for each of the discontinuous transitions within one period (three, in the case shown). Also separate continuous-time transition equations will be required for each different-duration interval between discontinuous transitions. Once such a set of equations has been determined it holds for *any* interval $[kT, kT + T)$.

If all the samplers operate periodically, but with different periods, we have the case of Fig. 7-9*c*. We assume that the various periods are in simple ratios to each other. Then the entire sampling pattern will repeat with a period T that represents the least common multiple of the sampling periods. A complete set of transition equations must be determined for an interval of length T. This set is then sufficient to describe the state at any time.

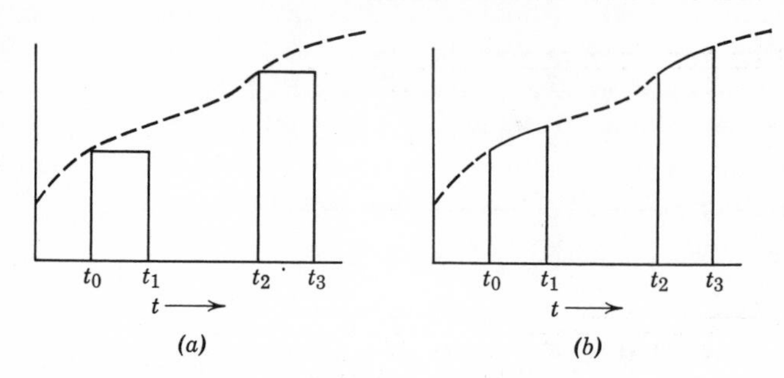

Figure 7-10 Fractional-period hold extrapolators.

In the case of 7-9*d*, the samplers are all synchronized; however, sampling occurs with a nonuniform pattern which repeats with a period T.[11] Transition equations are required for each of the discontinuous transitions and each of the continuous-transition intervals within one period of length T.

Various combinations of the foregoing types of sampling patterns are possible; in each case the required transition equations can be obtained in the manner already outlined. It is important to note that whenever the total sampling pattern is periodic with period T, we need derive transition equations for only *one* period.

If a sampled-data system contains hold extrapolators of first or higher order, we simply resolve these holds into components consisting of discrete-time subsystems, continuous-time subsystems, and zero-order holds,[12] and then proceed in the usual manner. Occasionally, a sampled-data system may contain hold-extrapolators of the *fractional-period* type, as shown in Fig. 7-10. Both holds of Fig. 7-10 have an output only during part of an interval. The hold of Fig. 7-10*a* is similar to a zero-order hold, except that it "holds" for only a fraction of an interval and has zero output for the remainder of the interval. The formulation of transition equations for a system containing one or more such extrapolators follows the procedure

[11] Cf. Section 4.6.

[12] Cf. Sections 5.7 and 5.8.

outlined in Section 7.4. The effect of having the outputs of these extrapolators abruptly go to zero at the end of the "holding" action is merely to introduce one or more additional discontinuous state transitions.

The fractional-period hold of Fig. 7-10*b* differs from any of the hold extrapolators previously considered in that, instead of sampling and holding, it transmits the continuous-time input to its output for a fraction of an interval and has zero output for the remainder of the interval. It is, of course, not a hold extrapolator at all, but merely a *switch* that closes for finite intervals of time. We shall refer to it as a *finite-duration sampler*. This kind of a system element has no memory; hence no state variable need be assigned to it. Its inputs must be continuous-time functions. No discontinuous state transitions are involved. However, the continuous-time transition equations for the system are, of course, different when the switch is "closed" than when it is "open."

7.7 SYSTEM STABILITY

The single, most important problem of system theory is the one associated with the concept of stability. Stability refers to the boundedness of the variables of a system as $t \to \infty$. Roughly speaking, a system subjected to a bounded input is considered stable if its state variables are of bounded variation and is considered unstable if they are not. However, when we attempt to formulate a precise definition of stability, we discover that we must distinguish among a number of different kinds of stability. We find that the system properties of linearity and stationarity have an important bearing on the definitions of stability that are applicable to a particular system.

There are two major points of view with respect to the stability of a system. In one, the system is presumed to possess an equilibrium state and our concern is with the ability of the system to maintain a state in the vicinity of this equilibrium in the absence of any inputs as $t \to \infty$. In the other, we assume that the system is subjected to a bounded input, and we focus attention on the resulting behavior of the state variables.

Our interest here lies solely with the stability of discrete-time and sampled-data systems. However, the development of the theory for these types of systems differs in only a few minor ways from that applicable to continuous-time systems. Also we shall make the tacit assumption that if a sampled-data system is stable at discrete instants t_k, $k = 0, 1, \ldots, \infty$, the system is stable for all time $t \geq t_0$. This assumption ignores the possibility that the state of a system may remain bounded at the discrete instants t_k and yet increase without bound in the intervals between successive instants.

Although such instability is theoretically possible, its occurrence in practice is too remote to deserve serious attention.[13]

7.8 STABILITY OF AN EQUILIBRIUM STATE

Let us consider a sampled-data system to which no input is applied.[14] We write

$$\mathbf{x}(t_{k+1}) = \mathbf{F}\big(\mathbf{x}(t_k); t_k\big) \tag{7.73}$$

where t_k is an independent, discrete time variable, $t_{k+1} > t_k$ for all integers k, $-\infty \leq k \leq \infty$, and $t_k \to \infty$ as $k \to \infty$. We shall denote the solution of (7.73) for an initial state $\mathbf{x}^0$ and an initial time t_0 by

$$\mathbf{x}(t_k) = \mathbf{P}(t_k; \mathbf{x}^0, t_0) \tag{7.74}$$

for all $t_k \geq t_0$. Further, we assume that for the systems under consideration the vector-valued function $\mathbf{P}$ is always continuous in all its arguments. Clearly,

$$\mathbf{P}(t_0; \mathbf{x}^0, t_0) = \mathbf{x}^0 \tag{7.75}$$

and

$$\mathbf{P}(t_a; \mathbf{x}^c, t_c) = \mathbf{P}\big(t_a; \mathbf{P}(t_b; \mathbf{x}^c, t_c), t_b\big) \tag{7.76}$$

for all $t_a \geq t_b \geq t_c$.

We now assume that

$$\mathbf{F}(\mathbf{0}; t_k) = \mathbf{0} \quad \text{for all } t_k \tag{7.77}$$

Then (7.73) possesses the trivial solution $\mathbf{x} = \mathbf{0}$. We shall refer to this solution as an *equilibrium state* of the system.[15] The equilibrium state $\mathbf{x} = \mathbf{0}$ is considered stable if the system starting from an initial state sufficiently close to $\mathbf{x} = \mathbf{0}$ remains in the neighborhood of $\mathbf{x} = \mathbf{0}$ as $t_k \to \infty$. To express this more precisely, we introduce the following definitions:

DEFINITION 1: The equilibrium state $\mathbf{x} = \mathbf{0}$ is said to be *L-stable* ("stable in the sense of Liapunov") if, for any t_0 and any $\varepsilon > 0$, there corresponds a $\delta(\varepsilon, t_0) > 0$ such that if $\|\mathbf{x}^0\| < \delta(\varepsilon, t_0)$, then $\|\mathbf{P}(t_k; \mathbf{x}^0, t_0)\| < \varepsilon$ for all $t_k \geq t_0$.[16]

[13] This assumes, of couse, that the instants t_k are independent of the state variables, as is the case for all sampled-data systems considered here.

[14] A system to which no input is applied is also referred to as a *free* system.

[15] An equilibrium state is any state $\mathbf{x}^e$ such that $\mathbf{F}(\mathbf{x}^e; t) = \mathbf{x}^e$ for all $t \geq t_0$.

[16] The symbol $\| \;\; \|$ denotes the Euclidean length (norm) of the enclosed vector. However, the definitions of stability given here hold for any suitable norm.

DEFINITION 2: The equilibrium state $\mathbf{x} = \mathbf{0}$ is said to be *uniformly L-stable* if, for any t_0 and any $\varepsilon > 0$, there corresponds a $\delta(\varepsilon) > 0$ such that if $\|\mathbf{x}^0\| < \delta(\varepsilon)$, then $\|\mathbf{P}(t_k; \mathbf{x}^0, t_0)\| < \varepsilon$ for all $t_k \geq t_0$.

DEFINITION 3: The equilibrium state $\mathbf{x} = \mathbf{0}$ is said to be *asymptotically stable* if it is L-stable and if there exists a $\eta(t_0) > 0$ such that

$$\lim_{k \to \infty} \|\mathbf{P}(t_k; \mathbf{x}^0, t_0)\| = 0 \tag{7.78}$$

for all $\|\mathbf{x}^0\| < \eta(t_0)$. If η is independent of t_0, the state is *uniformly asymptotically stable.*

Thus for an L-stable equilibrium state the system state remains in the vicinity of the equilibrium state, while for an asymptotically stable state the system state converges to the equilibrium state. If the conditions for these stabilities are independent of the initial time t_0, we indicate this by adding the term "uniform."

DEFINITION 4: The equilibrium state $\mathbf{x} = \mathbf{0}$ is said to be *asymptotically stable in the large* if it is asymptotically stable for *any* initial state $\mathbf{x}^0$.

We shall now develop means for determining whether a given system is stable in the sense of the foregoing definitions. However, before proceeding we must introduce the following additional definitions:

DEFINITION 5: A scalar function $V(\mathbf{x}; t_k)$ is said to be *positive definite* in a neighborhood $\mathcal{N}$ of the point $\mathbf{x} = \mathbf{0}$ if $V(\mathbf{0}; t_k) = 0$ and if there exists a continuous, nondecreasing, scalar function w such that

$$w(0) = 0 \tag{7.79}$$

and

$$V(\mathbf{x}; t_k) \geq w(\|\mathbf{x}\|) \tag{7.80}$$

for all t_k and all $\mathbf{x}$ in $\mathcal{N}$.

DEFINITION 6: A positive definite function $V(\mathbf{x}; t_k)$ is said to be *decrescent* in a neighborhood $\mathcal{N}$ if there exists a continuous nondecreasing scalar function s such that

$$s(0) = 0 \tag{7.81}$$

and

$$V(\mathbf{x}; t_k) \leq s(\|\mathbf{x}\|) \tag{7.82}$$

for all t_k and all $\mathbf{x} \neq \mathbf{0}$ in $\mathcal{N}$.

DEFINITION 7: A positive definite function $V(\mathbf{x}; t_k)$ is said to be *infinitely large* if $|V(\mathbf{x}; t_k)| \to \infty$ as $\|\mathbf{x}\| \to \infty$ for all t_k.

We are now ready to introduce a set of important stability theorems due to Liapunov.[17] Consider the system

$$\mathbf{x}(t_{k+1}) = \mathbf{F}(\mathbf{x}(t_k); t_k) \tag{7.73}$$

for which $\mathbf{x} = \mathbf{0}$ is an equilibrium state in accordance with (7.77). Let $V(\mathbf{x}; t_k)$ be a positive definite function with continuous first partial derivatives with respect to x. Let $\Delta V(\mathbf{x}; t_k)$ denote the first forward difference in $V(\mathbf{x}; t_k)$ along the positive time axis, that is,

$$\Delta V(\mathbf{x}; t_k) = \frac{V(\mathbf{x}(t_{k+1}); t_{k+1}) - V(\mathbf{x}(t_k); t_k)}{t_{k+1} - t_k} \tag{7.83}$$

Then

LIAPUNOV'S STABILITY THEOREM. *The equilibrium $\mathbf{x} = \mathbf{0}$ is L-stable if there exists a positive definite function $V(\mathbf{x}; t_k)$ possessing a nonpositive forward difference $\Delta V(\mathbf{x}; t_k)$.*

The theorem may be proved as follows. We refer to definitions (1) and (5). Given a particular $\varepsilon > 0$, we select a $\delta(\varepsilon, t_0) > 0$ such that for $\|\mathbf{x}^0\| < \delta(\varepsilon, t_0)$ we obtain both $\|\mathbf{x}^0\| < \varepsilon$ and $V(\mathbf{x}^0; t_0) < w(\varepsilon)$. Such a choice of δ is possible because of the continuity in $\mathbf{x}$ of $V(\mathbf{x}^0; t_k)$. Since $\Delta V(\mathbf{x}; t_k)$ is nonpositive,

$$V(\mathbf{x}^0; t_0) \geq V(\mathbf{P}(t_k; \mathbf{x}^0, t_0); t_k) \tag{7.84}$$

$$\geq w(\|\mathbf{P}(t_k; \mathbf{x}^0, t_0)\|) \tag{7.85}$$

and therefore

$$w(\varepsilon) > V(\mathbf{x}^0; t_0) \geq w(\|\mathbf{P}(t_k; \mathbf{x}^0, t_0)\|) \tag{7.86}$$

Since, however, w is a nondecreasing function, it follows that

$$\|\mathbf{P}(t_k; \mathbf{x}^0, t_0)\| < \varepsilon \tag{7.87}$$

for all $t_k \geq t_0$ and all $\|\mathbf{x}^0\| < \delta(\varepsilon, t_0)$.

LIAPUNOV'S ASYMPTOTIC STABILITY THEOREM. *The equilibrium $\mathbf{x} = \mathbf{0}$ is asymptotically stable if there exists a decrescent positive definite function $V(\mathbf{x}; t_k)$ possessing a negative definite forward difference $\Delta V(\mathbf{x}; t_k)$.*

The proof of this theorem is as follows.[18] From the proof of the L-stability theorem we know that the positive definite function $V(\mathbf{x}; t_k)$ has a non-negative limit as $t_k \to \infty$. We denote this limit by V_L. Since $V(\mathbf{x}; t_k)$

[17] A. Liapounoff, "Problème général de la stabilité du mouvement," originally published in 1892 in Russia, reprinted in *Annals of Mathematical Studies*, No. 17, Princeton University Press, Princeton, N.J., 1947.

[18] W. Hahn, "Über die Anwendung der Methode von Ljapunov auf Differenzengleichungen," *Math. Annalen*, vol. 136, pp. 430–441, 1958.

is decrescent by hypothesis,

$$V(\mathbf{x};\, t_k) \leq s(\|\mathbf{x}\|) \tag{7.88}$$

Hence $V_L > 0$ implies that the state magnitude $\|\mathbf{x}(t_k)\| = \|\mathbf{P}(t_k;\, \mathbf{x}^0,\, t_0)\|$ will always be larger than some positive number μ. Since $\Delta V(\mathbf{x};\, t_k)$ is negative definite,

$$\Delta V(\mathbf{x};\, t_k) \leq -r(\|\mathbf{x}\|) \tag{7.89}$$

where r is a continuous, nondecreasing scalar function, and where we have for simplicity assumed that $t_{k+1} - t_k = 1$. Then $V_L > 0$ implies

$$\Delta V(\mathbf{x};\, t_k) \leq -r(\mu) < 0 \tag{7.90}$$

We now write $V(\mathbf{x};\, t_k)$ in terms of its forward difference $\Delta V(\mathbf{x};\, t_k)$:

$$V(\mathbf{x}(t_k);\, t_k) = V(\mathbf{x}_0;\, t_0) + \sum_{i=0}^{k-1} \Delta V(\mathbf{x}(t_i);\, t_i) \tag{7.91}$$

$$\leq V(\mathbf{x}_0;\, t_0) - kr(\mu) \tag{7.92}$$

Since $V(\mathbf{x};\, t_k)$ is positive definite, the right-hand side of (7.92) may not become negative. The only way this can be satisfied for large k is to have $r(\mu) = 0$. Hence $\mu = 0$ and $\|\mathbf{P}(t_k;\, \mathbf{x}^0,\, t_0)\| \to 0$ as $k \to \infty$.

THEOREM ON ASYMPTOTIC STABILITY IN THE LARGE. *The equilibrium* $\mathbf{x} = \mathbf{0}$ *is asymptotically stable in the large if it is asymptotically stable and if* $V(\mathbf{x};\, t_k)$ *is such that* $w(\|\mathbf{x}\|) \to \infty$ *as* $\|\mathbf{x}\| \to \infty$.

The proof of this theorem follows directly from the proofs of the foregoing theorems.

We note that the stability (in the sense of Liapunov) of an equilibrium point depends on the existence of a positive definite function $V(\mathbf{x};\, t_k)$ possessing a nonpositive first difference. This function is commonly referred to as a Liapunov function. It is clear, of course, that this function will normally not be unique for a given equilibrium point.

The foregoing definitions and theorems hold for all systems that can be described by an equation of the form (7.73), whether linear or nonlinear, stationary or nonstationary. For stationary systems there is, of course, no distinction between stability and uniform stability. For linear systems (7.73) becomes simply

$$\mathbf{x}(t_{k+1}) = \mathbf{A}(t_k)\mathbf{x}(t_k) \tag{7.93}$$

The homogeneity of this equation implies that if the equilibrium state of a linear system is asymptotically stable, it is also asymptotically stable in the large.

7.9 STABILITY OF LINEAR, STATIONARY SYSTEMS

Let us consider the linear, stationary system described by the state equation

$$\mathbf{x}(k+1) = \mathbf{A}\mathbf{x}(k) \tag{7.94}$$

with a given arbitrary initial state $\mathbf{x}(0)$. If all the eigenvectors of $\mathbf{A}$ are independent, we can write

$$\mathbf{A} = \mathbf{H}\mathit{\Lambda}\mathbf{H}^{-1} \tag{7.95}$$

where

$$\mathit{\Lambda} = \begin{bmatrix} \lambda_1 & & & & 0 \\ & \lambda_2 & & & \\ & & \cdot & & \\ & & & \cdot & \\ & & & & \cdot \\ 0 & & & & \lambda_n \end{bmatrix} \tag{7.96}$$

and the λ_i are the eigenvalues of $\mathbf{A}$. From (2.28) the solution of (7.94) can then be written as

$$\mathbf{x}(k) = \mathbf{H}\mathit{\Lambda}^k\mathbf{H}^{-1}\mathbf{x}(0) \tag{7.97}$$

where

$$\mathit{\Lambda}^k = \begin{bmatrix} \lambda_1{}^k & & & & 0 \\ & \lambda_2{}^k & & & \\ & & \cdot & & \\ & & & \cdot & \\ & & & & \cdot \\ 0 & & & & \lambda_n{}^k \end{bmatrix} \tag{7.98}$$

It follows that $\lim_{k\to\infty} \|\mathbf{x}(k)\| = 0$ if $|\lambda_i| < 1$ for *all* $i = 1, 2, \ldots, n$, and that $\lim_{k\to\infty} \|\mathbf{x}(k)\| \to \infty$ if one or more $|\lambda_i| > 1$.

If $\mathbf{A}$ cannot be diagonalized, we make a transformation of variables

$$\mathbf{x} = \mathbf{K}\boldsymbol{\xi} \tag{7.99}$$

such that

$$\mathit{A} = \mathbf{K}^{-1}\mathbf{A}\mathbf{K} \tag{7.100}$$

is a *triangular* matrix. Then

$$\begin{bmatrix} \xi_1(k+1) \\ \xi_2(k+1) \\ \cdot \\ \cdot \\ \cdot \\ \xi_n(k+1) \end{bmatrix} = \begin{bmatrix} a_{11} & a_{12} & a_{13} & \cdots & a_{1n} \\ & a_{22} & a_{23} & \cdots & a_{2n} \\ & & \cdot & \cdots & \cdot \\ & & & \cdot & \cdot \\ & & & & \cdot \\ 0 & & & & a_{nn} \end{bmatrix} \begin{bmatrix} \xi_1(k) \\ \xi_2(k) \\ \cdot \\ \cdot \\ \cdot \\ \xi_n(k) \end{bmatrix} \tag{7.101}$$

or

$$\xi_i(k+1) = a_{ii}\xi_i(k) + \sum_{j=i+1}^{n} a_{ij}\xi_j(k), \quad i = 1, 2, \ldots, n \tag{7.102}$$

Now for $i = n$,

$$\xi_n(k+1) = a_{nn}\xi_n(k) \tag{7.103}$$

and hence for any $k \geq 0$ and any initial value $\boldsymbol{\xi}(0) = \mathbf{K}^{-1}\mathbf{x}(0)$,

$$\xi_n(k) = a_{nn}^k \xi_n(0) \tag{7.104}$$

Clearly, $\xi_n(k)$ will converge to zero as $k \to \infty$ if and only if $|a_{nn}| < 1$.

Let us now examine

$$\xi_{n-1}(k+1) = a_{n-1,n-1}\xi_{n-1}(k) + a_{n-1,n}\xi_n(k) \tag{7.105}$$

It is apparent from the same reasoning applied to (7.104) that $\xi_{n-1}(k)$ will converge to 0 as $k \to \infty$ if and only if $|a_{n-1,n-1}| < 1$ and simultaneously $\xi_n(k) \to 0$.

By induction it follows then from (7.102) that *all* $\xi_i(k) \to 0$ as $k \to \infty$ if and only if *all* $|a_{ii}| < 1$. Since the a_{ii} are the eigenvalues of A and since these are invariant under the transformation (7.100), we conclude that *a linear, stationary system*

$$\mathbf{x}(k+1) = \mathbf{A}\mathbf{x}(k)$$

is asymptotically stable if and only if all the eigenvalues of **A** *are of magnitude less than unity.*

The foregoing provides a necessary and sufficient condition for the stability of a linear, stationary system in terms of the eigenvalues of the unit-transition matrix **A**.

7.10 APPLICATION OF LIAPUNOV'S THEOREM

We can also develop a necessary and sufficient condition for stability in another way, namely by utilizing the concept of the Liapunov function. If we let E denote the forward-difference operator, then (7.94) may be written in the form

$$E\mathbf{x} = \mathbf{A}\mathbf{x} \tag{7.106}$$

We now select as a Liapunov function the positive definite quadratic form

$$V(\mathbf{x}) = \mathbf{x}'\mathbf{Q}\mathbf{x} \tag{7.107}$$

where the prime denotes the transpose of the vector and where **Q** is a symmetric constant matrix as yet to be determined. Let **R** be another symmetric, positive definite matrix such that

$$\Delta V(\mathbf{x}) = -\mathbf{x}'\mathbf{R}\mathbf{x} \tag{7.108}$$

Then

$$\Delta V(\mathbf{x}) = V(E\mathbf{x}) - V(\mathbf{x}) = V(\mathbf{A}\mathbf{x}) - V(\mathbf{x}) \tag{7.109}$$

Substituting (7.107) and making use of the fact that $(\mathbf{A}\mathbf{x})' = \mathbf{x}'\mathbf{A}'$, we find

$$\mathbf{x}'\mathbf{A}'\mathbf{Q}\mathbf{A}\mathbf{x} - \mathbf{x}'\mathbf{Q}\mathbf{x} = -\mathbf{x}'\mathbf{R}\mathbf{x} \tag{7.110}$$

and hence

$$\mathbf{A}'\mathbf{Q}\mathbf{A} - \mathbf{Q} = -\mathbf{R} \tag{7.111}$$

Now if we can solve (7.111) for the matrix $\mathbf{Q}$ and then show that $\mathbf{Q}$ is positive definite, the system (7.94) will be asymptotically stable in accordance with Liapunov's theorem.[19] To show that a solution of (7.111) exists we proceed as follows:[20]

From (7.110),

$$(\mathbf{A}\mathbf{x})'\mathbf{Q}\mathbf{A}\mathbf{x} - \mathbf{x}'\mathbf{Q}\mathbf{x} = -\mathbf{x}'\mathbf{R}\mathbf{x} \tag{7.112}$$

We let

$$M(\mathbf{x}) = \mathbf{x}'\mathbf{Q}\mathbf{x} \tag{7.113}$$

and

$$N(\mathbf{x}) = -\mathbf{x}'\mathbf{R}\mathbf{x} \tag{7.114}$$

Substituting in (7.112) gives

$$M(\mathbf{A}\mathbf{x}) - M(\mathbf{x}) = N(\mathbf{x}) \tag{7.115}$$

We next introduce a vector $\boldsymbol{\zeta}$ of dimension $\binom{n+1}{2}$ having components of the form

$$x_1^{i_1} x_2^{i_2} \ldots x_n^{i_n} \tag{7.116}$$

where the i_j take on all combinations of 0, 1, and 2 such that

$$i_1 + i_2 + \cdots + i_n = 2$$

For example, if $n = 2$ we would have

$$\boldsymbol{\zeta}(\mathbf{x}) = \begin{bmatrix} x_1^2 \\ x_1 x_2 \\ x_2^2 \end{bmatrix} \tag{7.117}$$

[19] Equation (7.111) is the discrete-time system counterpart of the equation $\mathbf{A}'\mathbf{Q} + \mathbf{Q}\mathbf{A} = -\mathbf{R}$ that holds for continuous-time systems; see, for example, R. Bellman, *Introduction to Matrix Analysis*, McGraw-Hill Book Co., New York, 1960, Chapter 13.

[20] W. Hahn, *op. cit.*

We can now express $M(\mathbf{x})$ and $N(\mathbf{x})$ in terms of scalar products involving the vector $\boldsymbol{\zeta}$. Thus we write

$$M(\mathbf{x}) = \mathbf{e}'\boldsymbol{\zeta}(\mathbf{x}) \tag{7.118}$$

$$N(\mathbf{x}) = \mathbf{f}'\boldsymbol{\zeta}(\mathbf{x}) \tag{7.119}$$

where $\mathbf{e}'$ and $\mathbf{f}'$ are row vectors. Then

$$M(\mathbf{Ax}) = \mathbf{e}'\boldsymbol{\zeta}(\mathbf{Ax}) = \mathbf{e}'\Gamma\boldsymbol{\zeta}(\mathbf{x}) \tag{7.120}$$

where Γ is a nonsingular transformation. Substituting in (7.115), we obtain

$$\mathbf{e}'(\Gamma - \mathbf{I})\boldsymbol{\zeta}(\mathbf{x}) = \mathbf{f}'\boldsymbol{\zeta}(\mathbf{x}) \tag{7.121}$$

This equation has a unique solution $\mathbf{e}'$ provided the determinant $|\Gamma - \mathbf{I}|$ does not vanish. From (7.113), (7.117), and (7.118), the existence of a unique solution $\mathbf{e}'$ in turn establishes the existence of a unique solution $\mathbf{Q}$.

Now from (7.120), (7.118), and (7.117), the eigenvalues of Γ take the form of the products

$$\lambda_1^{p_1} \; \lambda_2^{p_2} \; \ldots \; \lambda_n^{p_n} \tag{7.122}$$

where the p_j are either 0, 1, or 2 and $p_1 + p_2 + \cdots + p_n = 2$, and where the λ_j are the eigenvalues of $\mathbf{A}$. It follows from (7.121) that none of the products (7.122) may equal unity if (7.121) is to have a unique solution. This is equivalent to stating that (7.111) has a unique solution $\mathbf{Q}$ provided the eigenvalues of $\mathbf{A}$ satisfy the condition

$$\lambda_i\lambda_j \neq 1 \quad \text{for all } i, j \leq n \tag{7.123}$$

To determine whether $\mathbf{Q}$ is positive definite we make use of a theorem from matrix theory[21] according to which *a matrix is positive definite if and only if all the principal minors of its determinant are positive;* that is, we must have

$$\begin{gathered} q_{11} > 0 \\ \begin{vmatrix} q_{11} & q_{12} \\ q_{12} & q_{22} \end{vmatrix} > 0 \\ \begin{vmatrix} q_{11} & q_{12} & q_{13} \\ q_{12} & q_{22} & q_{23} \\ q_{13} & q_{23} & q_{33} \end{vmatrix} > 0 \\ \text{etc.} \end{gathered} \tag{7.124}$$

[21] See, for example, R. Bellman, *op. cit.*, pp. 72–74.

From the uniqueness of the solution of (7.111) and from the condition for stability developed in Section 7.9, we conclude that *the equation*

$$\mathbf{A}'\mathbf{Q}\mathbf{A} - \mathbf{Q} = -\mathbf{R}$$

is satisfied by a positive definite matrix **Q** *if and only if all the eigenvalues of* **A** *are of magnitude less than unity.*

As an illustration of the foregoing consider the system defined by the linear constant-coefficient difference equation

$$y(k + 2) + a_1 y(k + 1) + a_2 y(k) = 0 \tag{7.125}$$

If we let $y(k) = x_1(k)$ and $y(k + 1) = x_2(k)$, we may express (7.125) by the state equation

$$\begin{bmatrix} x_1(k+1) \\ x_2(k+1) \end{bmatrix} = \begin{bmatrix} 0 & 1 \\ -a_2 & -a_1 \end{bmatrix} \begin{bmatrix} x_1(k) \\ x_2(k) \end{bmatrix} \tag{7.126}$$

To determine whether this system is stable we must solve (7.111) for the unknown symmetric matrix **Q** and then show that this matrix is positive definite. Using (7.111), we let $\mathbf{R} = \mathbf{I}$ (the unit matrix) and write

$$\begin{bmatrix} 0 & -a_2 \\ 1 & -a_1 \end{bmatrix} \begin{bmatrix} q_{11} & q_{12} \\ q_{12} & q_{22} \end{bmatrix} \begin{bmatrix} 0 & 1 \\ -a_2 & -a_1 \end{bmatrix} - \begin{bmatrix} q_{11} & q_{12} \\ q_{12} & q_{22} \end{bmatrix} = \begin{bmatrix} -1 & 0 \\ 0 & -1 \end{bmatrix} \tag{7.127}$$

After performing the indicated multiplications, we obtain three equations in the three unknowns q_{11}, q_{12}, and q_{22}. In matrix form

$$\begin{bmatrix} -1 & 0 & a_2^2 \\ 0 & -(1 + a_2) & a_1 a_2 \\ 1 & -2a_1 & a_1^2 - 1 \end{bmatrix} \begin{bmatrix} q_{11} \\ q_{12} \\ q_{22} \end{bmatrix} = \begin{bmatrix} -1 \\ 0 \\ -1 \end{bmatrix} \tag{7.128}$$

Let us denote the eigenvalues of the system by α_1 and α_2. Then

$$\begin{aligned} a_1 &= -(\alpha_1 + \alpha_2) \\ a_2 &= \alpha_1 \alpha_2 \end{aligned} \tag{7.129}$$

and

$$\begin{bmatrix} -1 & 0 & \alpha_1^2 \alpha_2^2 \\ 0 & -(1 + \alpha_1 \alpha_2) & -\alpha_1 \alpha_2 (\alpha_1 + \alpha_2) \\ 1 & 2(\alpha_1 + \alpha_2) & (\alpha_1 + \alpha_2)^2 - 1 \end{bmatrix} \begin{bmatrix} q_{11} \\ q_{12} \\ q_{22} \end{bmatrix} = \begin{bmatrix} -1 \\ 0 \\ -1 \end{bmatrix} \tag{7.130}$$

The determinant of the 3×3 matrix is found to be

$$\Delta = (\alpha_1^2 - 1)(\alpha_2^2 - 1)(\alpha_1 \alpha_2 - 1) \tag{7.131}$$

Hence if (7.127) is to have a unique solution, we must have

$$|\alpha_1| \neq 1$$
$$|\alpha_2| \neq 1 \qquad (7.132)$$
$$\alpha_1\alpha_2 \neq 1$$

This is, of course, in agreement with (7.123).

We assume that conditions (7.132) are satisfied and apply Cramer's rule:

$$q_{11} = \frac{(\alpha_1^2 - 1)(\alpha_2^2 - 1)(\alpha_1\alpha_2 - 1) - 2\alpha_1^2\alpha_2^2(\alpha_1\alpha_2 + 1)}{(\alpha_1^2 - 1)(\alpha_2^2 - 1)(\alpha_1\alpha_2 - 1)} \qquad (7.133)$$

$$q_{12} = \frac{2(\alpha_1 + \alpha_2)\alpha_1\alpha_2}{(\alpha_1^2 - 1)(\alpha_2^2 - 1)(\alpha_1\alpha_2 - 1)} \qquad (7.134)$$

$$q_{22} = \frac{-2(\alpha_1\alpha_2 + 1)}{(\alpha_1^2 - 1)(\alpha_2^2 - 1)(\alpha_1\alpha_2 - 1)} \qquad (7.135)$$

To determine whether the matrix **Q** thus obtained is positive definite we examine the principal minors of its determinant. Since the principal minor q_{22} is of simpler form than q_{11}, we write for the conditions for positive definiteness

$$q_{22} > 0 \qquad (7.136)$$

and

$$q_{11}q_{22} - q_{12}^2 > 0 \qquad (7.137)$$

We find

$$q_{11}q_{22} - q_{12}^2 = \frac{2(1 + \alpha_1^2\alpha_2^2)}{(\alpha_1^2 - 1)(\alpha_2^2 - 1)(\alpha_1\alpha_2 - 1)^2} \qquad (7.138)$$

To satisfy (7.137), we must have

$$(\alpha_1^2 - 1)(\alpha_2^2 - 1) > 0 \qquad (7.139)$$

This implies that either

1. $$|\alpha_1| > 1 \quad \text{and} \quad |\alpha_2| > 1 \qquad (7.140a)$$

or

2. $$|\alpha_1| < 1 \quad \text{and} \quad |\alpha_2| < 1 \qquad (7.140b)$$

We now examine (7.135). Using (7.139), we can reduce the condition (7.136) to

$$\frac{1 + \alpha_1\alpha_2}{1 - \alpha_1\alpha_2} > 0 \qquad (7.141)$$

This condition is satisfied only for

$$-1 < \alpha_1\alpha_2 < 1 \qquad (7.142)$$

Combining the results of (7.140) and (7.142), we find that the matrix $\mathbf{Q}$ is positive definite if and only if

$$\begin{aligned} |\alpha_1| &< 1 \\ |\alpha_2| &< 1 \end{aligned} \tag{7.143}$$

As was to be expected, this result is identical with the condition for stability developed in Section 7.9.

7.11 STABILITY IN THE PRESENCE OF AN INPUT

Thus far we have discussed stability only from the point of view of an equilibrium—that is, whether a system will return to (or stay in the neighborhood of) an equilibrium point if it is given an initial displacement from this point. In the case of an input-free, linear, stationary system we saw that a system is asymptotically stable if and only if all the eigenvectors of its unit-transition matrix are of magnitude less than unity. We shall now investigate the stability of linear, stationary systems that are not input-free.

Since we shall permit a nonzero input for all future time, we must modify our definition of stability to take into account a boundedness condition on the input. Roughly speaking, we shall regard a system as stable if a "bounded input produces a bounded output." To express this more precisely we introduce the following definition:

A linear system is stable if and only if at any time t_0, with the system in any initial state $\mathbf{x}(t_0)$, every input $\mathbf{u}$ that satisfies the condition $\|\mathbf{u}(t)\| < \eta_1$ for all $t_0 \leq t < \infty$, yields a state $\mathbf{x}$ and an output $\mathbf{y}$ such that $\|\mathbf{x}(t)\| < \eta_2$ and $\|\mathbf{y}(t)\| < \eta_3$ for all $t_0 \leq t < \infty$. (η_1, η_2, and η_3 are finite constants.)

This definition of stability is applicable to continuous-time and discrete-time systems, whether stationary or nonstationary.

In Section 2.4 we saw that if a discrete-time system is described by the state equations

$$\begin{aligned} \mathbf{x}(k+1) &= \mathbf{A}(k)\mathbf{x}(k) + \mathbf{B}(k)\mathbf{u}(k) \\ \mathbf{y}(k) &= \mathbf{C}(k)\mathbf{x}(k) \end{aligned} \tag{7.144}$$

the state at any time k is given by the system transition equation

$$\mathbf{x}(k) = \mathbf{\Phi}(k, 0)\mathbf{x}(0) + \sum_{j=0}^{k-1} \mathbf{\Phi}(k, j+1)\mathbf{B}(j)\mathbf{u}(j) \tag{7.145}$$

where

$$\mathbf{\Phi}(k, j) = \mathbf{A}(k-1)\mathbf{A}(k-2)\ldots\mathbf{A}(j+1)\mathbf{A}(j) \tag{7.146}$$

and

$$\mathbf{\Phi}(k, k) = \mathbf{I} \tag{7.147}$$

If the input is bounded, that is, if

$$\|\mathbf{u}(k)\| < \eta_1 \quad \text{for all } 0 \leq k < \infty \tag{7.148}$$

then we can write

$$\|\mathbf{x}(k)\| \leq \|\mathbf{\Phi}(k, 0)\| \, \|\mathbf{x}(0)\| + \eta_1 \sum_{j=0}^{k-1} \|\mathbf{\Phi}(k, j+1)\mathbf{B}(j)\| \tag{7.149}$$

$$\|\mathbf{y}(k)\| \leq \|\mathbf{C}(k)\| \, \|\mathbf{x}(k)\| \tag{7.150}$$

Now if $\|\mathbf{A}(k)\|$, $\|\mathbf{B}(k)\|$, and $\|\mathbf{C}(k)\|$ remain finite as $k \to \infty$, both $\|\mathbf{x}(k)\|$ and $\|\mathbf{y}(k)\|$ will be bounded provided both $\|\mathbf{\Phi}(k, 0)\|$ and

$$\sum_{j=0}^{k-1} \|\mathbf{\Phi}(k, j+1)\mathbf{B}(j)\|$$

remain bounded.[22] The latter condition will be satisfied if two positive constants a_1 and a_2 exist such that for all k and j, $k \geq j$,

$$\|\mathbf{\Phi}(k, j)\| \leq a_1 e^{-a_2(k-j)} \tag{7.151}$$

For a *stationary* discrete-time system, the system matrices in (7.144) do not depend on k, and we have (from Section 2.4)

$$\mathbf{\Phi}(k, j) = \mathbf{A}^{k-j} \tag{7.152}$$

If we denote the Jordan canonical form of $\mathbf{A}$ by $\mathscr{D}$, we can write

$$\mathbf{\Phi}(k, j) = \mathbf{Q}\, \mathscr{D}^{k-j} \mathbf{Q}^{-1} \tag{7.153}$$

where $\mathbf{Q}$ is the required transformation.[23] Utilizing the result of Section 7.9, we observe that if $\mathbf{A}^{k-j}$ is to satisfy condition (7.151), all the eigenvalues of $\mathbf{A}$ must be of magnitudes less than unity. *We conclude that a stationary linear system subjected to a bounded input is stable if and only if all the zeros of the characteristic polynomial* $|\mathbf{A} - \lambda\mathbf{I}|$ *be within the circle* $|\lambda| = 1$ *in the complex* λ *plane.*

7.12 REDUCIBLE NONSTATIONARY AND NONLINEAR SYSTEMS

In Section 7.10 we described a systematic procedure for constructing a Liapunov function for a linear, stationary system. We shall now show

[22] The quantities $\|\mathbf{A}(k)\|$, etc., are *norms* of the indicated matrices. These norms are taken here in the following sense: $\|\mathbf{A}(k)\|$ is the supremum (i.e., least upper bound) of the Euclidean norm $\|\mathbf{A}(k)\mathbf{x}(k)\|$ for all $\mathbf{x}(k)$ satisfying $\|\mathbf{x}(k)\| = 1$. (See, e.g., R. Bellman, *op. cit.*, p. 162.)

[23] Cf. Section 2.7.

that this procedure can be applied also for determining the stability of certain classes of nonstationary and nonlinear systems.

Reducible Nonstationary Systems

We first consider the linear, nonstationary system represented by the state equation

$$\mathbf{x}(k+1) = \mathbf{A}(k)\mathbf{x}(k) \tag{7.154}$$

Let $\mathbf{S}(k)$ be a $n \times n$ matrix with bounded coefficients that is nonsingular for all $k \geq 0$. Let

$$\tilde{\mathbf{x}}(k+1) = \tilde{\mathbf{A}}(k)\tilde{\mathbf{x}}(k) \tag{7.155}$$

be the state equation of a system related to the first system in accordance with

$$\tilde{\mathbf{x}}(k) = \mathbf{S}(k)\mathbf{x}(k) \tag{7.156}$$

for all $k \geq 0$. Upon combining (7.156) and (7.154), we obtain

$$\tilde{\mathbf{x}}(k+1) = \mathbf{S}(k+1)\mathbf{A}(k)\mathbf{S}^{-1}(k)\tilde{\mathbf{x}}(k) \tag{7.157}$$

From (7.155),

$$\tilde{\mathbf{A}}(k) = \mathbf{S}(k+1)\mathbf{A}(k)\mathbf{S}^{-1}(k) \tag{7.158}$$

If the nature of $\mathbf{S}(k)$ is as specified, the system corresponding to (7.157) has precisely the same stability properties as that corresponding to (7.154). This fact can be used to advantage in establishing the stability of certain nonstationary systems.[24]

Let us suppose that for a given linear, nonstationary system (7.154), a bounded, nonsingular transformation $\mathbf{S}(k)$ exists such that

$$\mathbf{S}(k+1)\mathbf{A}(k)\mathbf{S}^{-1}(k) = \tilde{\mathbf{A}} = \text{constant} \tag{7.159}$$

for all $k \geq 0$. We can then determine the stability of the given *nonstationary* system by studying the stability of the transformed *stationary* system

$$\tilde{\mathbf{x}}(k+1) = \tilde{\mathbf{A}}\tilde{\mathbf{x}}(k) \tag{7.160}$$

using the method of Section 7.10. Systems that can be transformed in this manner into stationary systems are known as *reducible* systems.

Reducible Nonlinear Systems

Let us now consider the nonlinear, stationary system represented by

$$\mathbf{x}(k+1) = \mathbf{A}\mathbf{x}(k) + \mathcal{K}(\mathbf{x}(k)) \tag{7.161}$$

[24] A. Liapounoff, *op. cit.*

where $\mathcal{K}(\mathbf{x}(k))$ is a nonlinear vector-valued function of $\mathbf{x}(k)$. We impose the restriction that the components of $\mathcal{K}$ be representable by power series in the components x_i of the vector $\mathbf{x}$. If we further require that the power series involve terms of only second or higher order in the x_i, then in the neighborhood of the origin, $\mathbf{x} = \mathbf{0}$, the stability properties of the system (7.161) will be the same as those of the system

$$\mathbf{x}(k+1) = \mathbf{A}\mathbf{x}(k) \tag{7.162}$$

The latter system is referred to as the *first approximation* of (7.161).[25]

As an illustration consider the 2 × 2 system described by the equations

$$x_1(k+1) = \tfrac{1}{2}x_1(k) + \tfrac{1}{3}x_2(k) + 5x_1^2(k) + 3x_1^2(k)x_2^2(k)$$
$$x_2(k+1) = \tfrac{1}{3}x_1(k) + \tfrac{1}{4}x_2(k) + \tfrac{1}{2}x_2^2(k)$$

In accordance with the foregoing, the stability condition of the equilibrium point $\mathbf{x} = \mathbf{0}$ is the same as in the case of the system (7.162) with

$$\mathbf{A} = \begin{bmatrix} \frac{1}{2} & \frac{1}{3} \\ \frac{1}{3} & \frac{1}{4} \end{bmatrix}$$

Note that we can, of course, draw no conclusions about the stability "in-the-large" of the system from the first approximation.[26]

7.13 STABILITY CRITERIA

We have seen that for all stationary linear systems, as well as for non-stationary or nonlinear systems that are reducible to stationary linear systems, the necessary and sufficient condition for asymptotic stability is that the magnitudes of the eigenvalues must be less than unity. To determine the eigenvalues of an nth order system, we must solve the nth-degree characteristic equation $|\mathbf{A} - \lambda\mathbf{I}| = 0$. Without the availability of a digital computer, solving for the eigenvalues may be quite laborious if $n > 3$. Fortunately, a number of tests exist that enable us to determine whether or not all the roots of a characteristic equation are of magnitude less than unity without requiring us to solve for the roots themselves. Such tests are known as *stability criteria*.

[25] A. Liapounoff, *op. cit.*

[26] For a somewhat more general result see O. Perron, "Über Stabilität und asymptotisches Verhalten der Lösungen eines Systems endlicher Differenzengleichungen," *J. reine angew. Math.*, vol. 161, pp. 41–64, 1929.

Modified Schur-Cohn Criterion

The modified Schur-Cohn criterion utilizes a simple numerical procedure to determine whether a given polynomial with real coefficients,

$$F(z) = a_n z^n + a_{n-1} z^{n-1} + \cdots + a_1 z + a_0, \qquad a_n > 0 \qquad (7.163)$$

has *all* its zeros in the interior of the unit circle. The criterion was introduced by Tsypkin[27] and Jury[28] and represents a considerable simplification of an older criterion due to Schur-Cohn.[29]

Given the polynomial (7.163), with $a_n > 0$, we define its *inverse polynomial*:

$$F^{-1}(z) = z^n F(z^{-1}) \qquad (7.164)$$

$$= a_0 z^n + a_1 z^{n-1} + \cdots + a_{n-1} z + a_n \qquad (7.165)$$

The roots of $F^{-1}(z)$ are the inverses of the roots of $F(z)$ with respect to the circle $|z| = 1$. In addition,

$$[F^{-1}(z)]^{-1} = F(z) \qquad (7.166)$$

With the polynomials in the form of (7.163) and (7.165), we divide $F^{-1}(z)$ by $F(z)$, beginning at the left (high-power) end, to obtain one quotient term and a remainder:

$$\frac{F^{-1}(z)}{F(z)} = \alpha_0 + \frac{F_1^{-1}(z)}{F(z)} \qquad (7.167)$$

The remainder $F_1^{-1}(z)$ will be a polynomial of degree $n - 1$. The quotient term α_0 is equal to a_0/a_n.

The division is now repeated using the remainder polynomial $F_1^{-1}(z)$ and its inverse polynomial $F_1(z)$ in accordance with the recursive relation

$$\boxed{\frac{F_i^{-1}(z)}{F_i(z)} = \alpha_i + \frac{F_{i+1}^{-1}(z)}{F_i(z)}, \qquad i = 0, 1, \ldots, n-2} \qquad (7.168)$$

where, of course, $F_0(z) = F(z)$.

The necessary and sufficient test that the roots of the equation $F(z) = 0$ lie in the interior of the unit circle in the z plane is that *all* the following

[27] Y. Z. Tsypkin, *Theory of Impulse Systems* (in Russian), State Publisher for Physical-Mathematical Literature, Moscow, pp. 423–428, 1958.

[28] E. I. Jury, "A Stability Test for Linear Discrete Systems Using a Simple Division," *Proc. IRE*, vol. 49, no. 12 (correspondence section), p. 1948, December 1961.

[29] M. Marden, *The Geometry of the Zeros of a Polynomial in a Complex Plane*, American Mathematical Society, New York, pp. 148–155, 1949.

three conditions be satisfied:

$$
\begin{aligned}
&(a) \qquad && F(1) > 0 && \\
&(b) && F(-1) < 0 && \text{for } n \text{ odd} \\
& && > 0 && \text{for } n \text{ even} \\
&(c) && |\alpha_i| < 1, && i = 0, 1, \ldots, n-2
\end{aligned}
\tag{7.169}
$$

Because of their simplicity, conditions (*a*) and (*b*) are examined first. If they are not satisfied, it is already established that $F(z) = 0$ has at least one root that is not inside the unit circle. There is then no need to examine condition (*c*).

The foregoing can be applied directly to determining whether the characteristic equation $|\mathbf{A} - \lambda\mathbf{I}| = 0$ has all its roots inside the unit circle provided the left-hand side of this equation can be written in the form of a polynomial. If a discrete-time system is represented by a z-transfer function, its characteristic equation is simply the denominator of the transfer function.

Example 1

We wish to determine whether a system having the following characteristic equation is stable:

$$|\mathbf{A} - \lambda\mathbf{I}| = 10\lambda^3 - 41\lambda^2 + 54\lambda - 5 = 0$$

Letting $F(\lambda) = |\mathbf{A} - \lambda\mathbf{I}|$, we find

$$
\begin{aligned}
&(a) \qquad && F(1) = 18 > 0 \\
&(b) && F(-1) = -110 < 0, \qquad (n = 3, \text{odd})
\end{aligned}
$$

Since both conditions are satisfied, we must examine condition (*c*):

$$
(c) \qquad
\begin{aligned}
F^{-1}(\lambda) &= -5\lambda^3 + 54\lambda^2 - 41\lambda + 10 \\
\frac{F^{-1}(\lambda)}{F(\lambda)} &= -0.5 + \frac{33.5\lambda^2 - 14\lambda + 7.5}{10\lambda^3 - 41\lambda^2 + 54\lambda - 5} \\
\frac{F_1^{-1}(\lambda)}{F_1(\lambda)} &= \frac{33.5}{7.5} + \frac{48.5\lambda - 142.3}{7.5\lambda^2 - 14\lambda + 33.5}
\end{aligned}
$$

Since $n = 3$, we need only determine α_0 and α_1,

$$|\alpha_0| = |0.5| < 1$$

$$|\alpha_1| = \left|\frac{33.5}{7.5}\right| > 1 \text{ (violated)}$$

Hence the system is unstable. (The actual roots are $\lambda = 2 - j$, $2 + j$, and 0.1.)

If $F(z)$ has z as a factor, we obtain $\alpha_0 = 0$, which, of course, satisfies condition (*c*). A zero α-value can also be obtained when in performing the division (7.168), we obtain a remainder that is *two* degrees lower in z than $F_i^{-1}(z)$. Neither case presents any difficulty. We simply continue to use (7.168) until *all* the necessary α_i, $i = 0, 1, \ldots, n-2$, are obtained.

Example 2

Determine whether the system described by the transfer function

$$H(z) = \frac{2z^2 - 3z + 1}{8z^4 + 4z^3 + 2z^2 + 4z}$$

is stable.

We write

$$F(z) = 8z^4 + 4z^3 + 2z^2 + 4z$$
$$F^{-1}(z) = 4z^3 + 2z^2 + 4z + 8$$
$$(a)\quad F(1) = 18, \qquad (b)\quad F(-1) = 2$$

Conditions (*a*) and (*b*) are satisfied; hence we proceed to (*c*).

$$(c)\qquad \frac{F^{-1}(z)}{F(z)} = 0 + \frac{4z^3 + 2z^2 + 4z + 8}{8z^4 + 4z^3 + 2z^2 + 4z}, \qquad \alpha_0 = 0$$
$$\frac{F_1^{-1}(z)}{F_1(z)} = \frac{1}{2} + \frac{3z + 6}{8z^3 + 4z^2 + 2z + 4}, \qquad \alpha_1 = \tfrac{1}{2}$$
$$\alpha_2 = 0$$

Since all the α_i, $i = 0, 1, 2$, satisfy condition (*c*), all the zeros of $F(z)$ lie inside the circle $|z| = 1$.

Other Stability Criteria

In addition to the modified Schur-Cohn criterion, we may also employ the Routh-Hurwitz,[30] Liénard-Chipart,[31] or Nyquist[32] criteria. Since these criteria are designed to test for the stability of continuous-time systems, they require some modification before they can be used for discrete-time systems.

The Routh-Hurwitz and Liénard-Chipart criteria are algebraic tests for determining whether a given polynomial has all its zeros in the left-half complex plane. To apply these tests to a discrete-time system, it is first necessary to map the interior of the unit circle in the complex z plane into

[30] See, for example, F. R. Gantmacher, *Applications of the Theory of Matrices*, Interscience Publishers, New York, 1959, Chapter 5.

[31] *Ibid.*, pp. 262–268.

[32] See, for example, R. G. Brown and J. W. Nilsson, *Introduction to Linear Systems Analysis*, John Wiley and Sons, New York, 1962.

the left half of a complex w plane. This can be accomplished by means of the bilinear transformation

$$z = \frac{w + 1}{w - 1} \tag{7.170}$$

Thus if we wish to know whether a characteristic polynomial such as (7.163) has all its zeros inside the circle $|z| = 1$, we first make the substitution (7.170) and then apply either the Routh-Hurwitz or the Liénard-Chipart test to the numerator polynomial of the resulting function of w. If the latter polynomial has all its zeros in the left-half w plane, the former will have all its zeros inside $|z| = 1$.

The Nyquist stability criterion is based on a mapping theorem of Cauchy. If a complex variable z in the z plane describes a closed contour C_1 in a positive sense, then $H(z)$, a function of the complex variable z, will describe a closed contour C_2 in the $H(z)$ plane and will encircle the origin η times in the positive direction, where η is the difference between the number of zeros and poles of $H(z)$ enclosed by C_1. Multiple poles or zeros must be counted according to their multiplicity.

The use of the Nyquist criterion for discrete-time systems is essentially the same as for continuous-time systems except that instead of encircling the right-half complex s plane we must now encircle the exterior of the unit circle in the z plane.

When compared with the modified Schur-Cohn criterion, neither the Routh-Hurwitz, Liénard-Chipart, nor Nyquist criteria appear attractive for use with discrete-time systems. The Routh-Hurwitz and Liénard-Chipart criteria are cumbersome because of the need to apply the bilinear transformation, and the Nyquist criterion requires a tedious graphical construction.

REFERENCES

Ash, R., W. H. Kim, and G. M. Kranc, "A General Flow Graph Technique for the Solution of Multiloop Sampled Systems," *Trans. ASME, J. Basic Engrg.*, vol. 82, pp. 360–370, June 1960.

Barker, R. H., "The Pulse Transfer Function and Its Application to Sampling Servo Systems," *Proc. IEE*, vol. 99, part 4, pp, 302–317, London, December 1952.

Bellman, R., *Introduction to Matrix Analysis*, McGraw-Hill Book Co., New York, 1960.

Bellman, R., "Kronecker Products and the Second Method of Lyapunov," *Math. Nachrichten*, vol. 20, pp. 17–19, 1959.

Chetayev, N. G., *The Stability of Motion*, Pergamon Press, New York, 1961.

Coates, C. L., "Flowgraph Solutions of Linear Algebraic Equations, "*IRE Trans. Circuit Theory*, vol. CT-6, pp. 170–187, June 1959.

Desoer, C. A., "The Optimum Formula for the Gain of a Flow Graph or a Simple Derivation of Coates' Formula," *Proc. IRE*, vol. 48, pp. 883–889, May 1960.

Friedland, B., O. Wing, and R. Ash, *Principles of Linear Networks*, McGraw-Hill Book Co., New York, 1961.

Gantmacher, F. R., *Applications of the Theory of Matrices*, Interscience Publishers, New York, 1959.

Hahn, W., "Eine Bemerkung zur Zweiten Methode von Ljapunov," *Math. Nachrichten*, vol. 14, pp. 349–354, 1956.

Hahn, W., *Theory and Application of Liapunov's Direct Method*, Prentice-Hall, Englewood Cliffs, N.J., 1963.

Hahn, W., "Über die Anwendung der Methode von Ljapunov auf Differenzengleichungen," *Math. Annalen*, vol. 136, pp. 430–441, 1958.

Jury, E. I., "On the Evaluation of the Stability Determinants in Linear Discrete Systems," *IRE Trans. Autom. Control*, vol. AC-7, no. 4, pp. 51–55, July 1962.

Jury, E. I. *Sampled-Data Control Systems*, McGraw-Hill Book Co., New York, 1958.

Jury, E. I., "A Stability Test for Linear Discrete Systems Using a Simple Division," *Proc. IRE*, vol. 49, no. 12, pp. 1948–1949, December 1961.

Jury, E. I. and B. H. Bharucha, "Notes on the Stability Criterion for Linear Discrete Systems," *IRE Trans. Autom. Control*, vol. AC-6, no. 1, pp. 88–90, February 1961.

Jury, E. I. and J. Blanchard, "A Stability Test for Linear Discrete Systems in Table Form," *Proc. IRE*, vol. 49, no. 12, pp. 1947–1948, December 1961.

Kalman, R., "On the Stability of Time-Varying Linear Systems," *IRE Trans. Circuit Theory*, vol. CT-9, no. 4, pp. 420–422, December 1962.

Kalman, R. and J. Bertram, "Control System Analysis and Design via the 'Second Method' of Lyapunov," *Trans. ASME, J. Basic Engrg.*, vol. 82, Ser. D: I, Continuous-Time Systems, pp. 371–393; II, Discrete-Time Systems, pp. 394–399; 1960.

Kalman, R. and J. Bertram, "A Unified Approach to the Theory of Sampling Systems," *J. Franklin Inst.*, vol. 267, pp. 405–436, May 1959.

Krasovskii, N. N., *Stability of Motion*, Stanford University Press, Stanford, Calif., 1963.

Kuo, Benjamin, *Analysis and Synthesis of Sampled-Data Control Systems*, Prentice-Hall, Englewood Cliffs, N.J., 1963.

LaSalle, J. P., "Stability and Control," *J. Soc. Ind. Appl. Math., Ser. A: Control*, vol. 1, no. 1, pp. 3–15, 1962.

LaSalle, J. P. and S. Lefschetz, *Stability by Liapunov's Direct Method, with Applications*, Academic Press, New York, 1961.

Lendaris, G. G., "Input-Output Relationships for Multisampled Loop Systems," *Proc. IRE*, vol. 49, no. 11, November 1961.

Lendaris, G. G. and E. I. Jury, "Input-Output Relationships for Multisampled-Loop Systems," *Trans. AIEE*, vol. 78, part II, pp. 375–385, January 1960.

Leondes, C. T. (ed.), *Computer Control Systems Technology*, McGraw-Hill Book Co., New York, 1961.

Li, Ta, "Die Stabilitätsfrage bei Differenzengleichungen," *Acta Math.*, vol. 63, pp. 99–141, 1934.

Liapounoff, A., "Problème général de la stabilité du mouvement," original published in 1892 in Russia, reprinted in *Annals of Mathematical Studies*, no. 17, Princeton University Press, Princeton, N.J., 1947.

Malkin, I. G., *Theory of Stability of Motion* (English translation), AEC-tr-3352, Office of Technical Services, U.S. Dept. of Commerce, Washington, D.C.

Marden, M., *The Geometry of the Zeros of a Polynomial in a Complex Variable*, American Mathematical Society, New York, 1949.

Mason, S. J., "Feedback Theory—Further Properties of Signal Flow Graphs," *Proc. IRE*, vol. 44, no. 7, pp. 920–926, July 1956.

Monroe, A. J., *Digital Processes for Sampled Data Systems*, John Wiley and Sons, New York, 1962.

Perron, O., "Über die Poincaresche lineare Differenzgleichung," *J. reine angew. Math.*, vol. 137, pp. 6–64, 1910.

Perron, O., "Über Stabilität und asymptotisches Verhalten der Lösungen eines Systems endlicher Differenzengleichungen," *J. reine angew. Math.*, vol. 161, pp. 41–64, 1929.

Poincaré, H., "Sur les equations lineaires aux differentielles ordinaires et aux differences finies," *Am. J., Math.*, vol. 7, pp. 203–258, 1885.

Ragazzini, J. R. and G. Franklin, *Sampled-Data Control Systems*, McGraw-Hill Book Co., New York, 1958.

Ralston, A., "A Symmetric Matrix Formulation of the Hurwitz-Routh Stability Criterion," *IRE Trans. Autom. Control*, vol. AC-7, no. 4, pp. 50–51, July 1962.

Salzer, J. M., " Signal Flow Reductions in Sampled-Data Systems," *IRE Wescon Conv. Rec.*, part 4, pp. 166–170, 1957.

Schur, J., "Über Potenzreihen die im Innern des Einheitskreises beschränkt sind," *J. reine angew. Math.*, vol. 147, pp. 205–232, 1916, and vol. 148, pp. 112–145, 1917.

Tou, J. T., *Digital and Sampled-Data Control Systems*, McGraw-Hill Book Co., New York, 1959.

Tsypkin, Y. Z., *Theory of Impulse Systems*, State Publisher for Physical-Mathematical Literature, Moscow, 1958.

Wilf, H. S., "A Stability Criterion for Numerical Integration," *J. Assoc. Comput. Mach.*, vol. 6, no. 3, pp. 363–365, July 1959.

CHAPTER 8

Discrete Stochastic Processes

8.1 INTRODUCTION

We must distinguish between time sequences that are *deterministic* and those that are *stochastic*. A time sequence is considered to be deterministic if the future values that it may assume can be precisely predicted, that is, without any element of chance or uncertainty. In contrast, a stochastic time sequence is one for which future values can be predicted only in a statistical sense. We find it convenient to define a *discrete stochastic process* as an ensemble of stochastic time sequences $\{f(k)\}$ that can be characterized by a set of probability density functions of all orders.[1]

The behavior of physical phenomena is, of course, never *precisely* predictable. Depending on the nature of the phenomenon, the degree to which the phenomenon is understood, the precision with which its behavior is to be described, as well as the purpose of the description, we may select either a deterministic or a stochastic variable to represent the particular behavior characteristic. In effect, we may decide whether, for the objectives at hand, it is preferable to give the description in the form of a variable defined in rigid, uncertainty-free terms, or to give it in terms of its statistical properties. Both such descriptions involve approximations; sometimes one is to be preferred, sometimes the other.

In the preceding chapters we have been concerned only with deterministic behavior. In the present chapter we shall apply some of the techniques of discrete-time system theory to the study of stochastic processes. Of particular interest are methods for the solution of system problems where the inputs and outputs are stochastic processes.

A stochastic process is said to be *stationary* if its statistical properties remain constant with time. Most of the commonly encountered stationary processes obey the so-called *ergodic hypothesis*; that is, the statistical

[1] For a detailed discussion of discrete stochastic processes, see Y. W. Lee, *Statistical Theory of Communication*, John Wiley and Sons, New York, 1960; W. B. Davenport and W. L. Root, *An Introduction to the Theory of Random Signals and Noise*, McGraw-Hill Book Co., New York, 1958; and J. L. Doob, *Stochastic Processes*, John Wiley and Sons, New York, 1953.

properties obtained by averaging one time sequence of the process over all time are identical to those obtained by averaging over the ensemble of time sequences at any particular instant of time.

8.2 DEFINITIONS OF BASIC TERMS

Consider a discrete stochastic process of M sequences. Let μ_1 be the fraction of the sequences which at the argument value k have an amplitude lying between x and $x + \Delta x$. Then the *first-order probability density* is defined as

$$p_1(x, k) = \lim_{\substack{M \to \infty \\ \Delta x \to 0}} \frac{\mu_1}{\Delta x} \tag{8.1}$$

Clearly,

$$\int_{-\infty}^{\infty} p_1(x, k)\, dx = 1 \tag{8.2}$$

If we let μ_2 be the fraction of sequences which at time k_1 have an amplitude between x_1 and $x_1 + \Delta x_1$, and in addition have at time k_2 an amplitude between x_2 and $x_2 + \Delta x_2$, then the *second-order probability density* is given by

$$p_2(x_1, k_1; x_2, k_2) = \lim_{\substack{M \to \infty \\ \Delta x_1 \to 0 \\ \Delta x_2 \to 0}} \frac{\mu_2}{\Delta x_1\, \Delta x_2} \tag{8.3}$$

From the definition of the second-order probability density it follows that

$$\int_{-\infty}^{\infty} p_2(x_1, k_1; x_2, k_2)\, dx_2 = p_1(x_1, k_1) \tag{8.4}$$

The foregoing can be extended to define probability densities of all orders. Clearly, the higher the order of a given probability density, the more that will be known about the stochastic process to which it refers. A probability density of order η can always be obtained from the corresponding probability density of order $\eta + 1$ since

$$p_\eta(x_1, k_1; \ldots; x_\eta, k_\eta) = \int_{-\infty}^{\infty} p_{\eta+1}(x_1, k_1; \ldots; x_{\eta+1}, k_{\eta+1})\, dx_{\eta+1} \tag{8.5}$$

In working with two or more stochastic processes we find it convenient to utilize so-called *joint probability densities*. Consider two stochastic processes X and Y that are related in the sense that each time sequence in X is uniquely associated with a time sequence in Y, and vice versa. Assume that we are given M pairs of sequences, each consisting of one sequence from X and its associated sequence from Y. Let μ be the fraction of these M pairs for which the sequence belonging to X lies between x and $x + \Delta x$ at k_1, and the sequence belonging to Y lies between y and $y + \Delta y$ at k_2, We then define the *joint probability density of order 1, 1* for X and Y as

$$p_{11}(x, k_1; y, k_2) = \lim_{\substack{M \to \infty \\ \Delta x \to 0 \\ \Delta y \to 0}} \frac{\mu}{\Delta x \, \Delta y} \tag{8.6}$$

If the processes X and Y are statistically independent, the joint probability density (8.6) will be equal to simply the product of the first-order probability densities of X and Y.

Higher-order joint probability densities are similarly defined. The joint probability densities provide a means for representing the dependencies existing among the different processes to which they refer.

We note that if a process is stationary, its probability densities are independent of the time origin. Thus $p_1(x, k)$ and $p_2(x_1, k_1; x_2, k_2)$ become simply $p_1(x)$ and $p_2(x_1, x_2, k_2 - k_1)$, respectively, in the stationary case.

A variety of important relations are obtained when moments of the probability densities are considered. For example, for a stationary process, the first moment

$$\tilde{x} = \int_{-\infty}^{\infty} x p_1(x) \, dx \tag{8.7}$$

represents the *mean* or *expected value* of the process. The second moment

$$\widetilde{x^2} = \int_{-\infty}^{\infty} x^2 p_1(x) \, dx \tag{8.8}$$

represents the *mean square value* of the process.

If the mean is subtracted from x in (8.8), we obtain the *second central moment* or *variance*, denoted by ${\sigma_x}^2$:

$${\sigma_x}^2 = \widetilde{(x - \tilde{x})^2} = \int_{-\infty}^{\infty} (x - \tilde{x})^2 p_1(x) \, dx \tag{8.9}$$

The quantity σ_x is also referred to as the *standard deviation* of the process.

The moment

$$R_x(k_1, k_2) = \int_{-\infty}^{\infty} dx_1 \int_{-\infty}^{\infty} dx_2 \, x_1 x_2 p_2(x_1, k_1; x_2, k_2) \tag{8.10}$$

is known as the *autocorrelation function* of the process. Similarly, the moment

$$R_{xy}(k_1, k_2) = \int_{-\infty}^{\infty} dx \int_{-\infty}^{\infty} dy \, xy p_{11}(x, k_1; y, k_2) \tag{8.11}$$

is called the *crosscorrelation function* for the two processes X and Y.

For stationary processes, the correlation functions are written simply as $R_x(l)$ and $R_{xy}(l)$, respectively, where $l = k_2 - k_1$.

The foregoing definitions are all based on ensemble averages, that is, on averages taken over the infinite set of sequences of the process at a particular argument value k. In the remaining sections of this chapter, only stationary stochastic processes will be considered and the ergodic hypothesis will be assumed to apply. Accordingly, the various moments of the preceding section can also be defined in terms of time averages.

For the *mean* value, we may write

$$\overline{x} = \lim_{K\to\infty} \frac{1}{2K+1} \sum_{k=-K}^{K} x(k) \tag{8.12}$$

where the overscoring denotes a time average. According to the ergodic hypothesis, we have from (8.7)

$$\bar{x} = \tilde{x}$$

Similarly, we have corresponding to the mean square value of (8.8)

$$\overline{x^2} = \lim_{K\to\infty} \frac{1}{2K+1} \sum_{k=-K}^{K} x^2(k) \tag{8.13}$$

Of especial importance are the correlation sequences corresponding to (8.10) and (8.11) for the case of discrete, stationary processes. Thus the *autocorrelation sequence* is given by

$$R_x(l) = \lim_{K\to\infty} \frac{1}{2K+1} \sum_{k=-K}^{K} x(k)x(k+l) \tag{8.14}$$

and the *crosscorrelation* sequence is given by

$$R_{xy}(l) = \lim_{K\to\infty} \frac{1}{2K+1} \sum_{k=-K}^{K} x(k)y(k+l) \tag{8.15}$$

In many practical cases of system analysis it has been found that the autocorrelation sequence provides a workable and satisfactory characterization of a stochastic process. Similarly, the crosscorrelation sequence generally serves as an acceptable measure of the coherence between two related processes. Both correlation sequences are relatively easy to obtain experimentally; hence they have found wide acceptance in system analysis.

It is easily shown[2] that the following properties apply to the correlation sequences:

$$R_x(0) = \overline{x^2(k)} \tag{8.16}$$

$$R_x(0) \geq R_x(l) \tag{8.17}$$

$$R_x(-l) = R_x(l) \tag{8.18}$$

$$R_{xy}(-l) = R_{yx}(l) \tag{8.19}$$

$$|R_{xy}(l)| \leq \sqrt{R_x(0)R_y(0)} \tag{8.20}$$

[2] See, for example, Y. W. Lee, *op. cit.*, pp. 51–78.

8.3 SYSTEM RELATIONS FOR THE CORRELATION SEQUENCES

Consider a linear, stationary, discrete-time system (which need not be physically realizable) that is described by the weighting sequence $\{h(k)\}$. Let the input and output be denoted by u and y respectively. Then from (2.7)

$$y(k) = \sum_{p=-\infty}^{\infty} h(p)u(k-p) \tag{8.21}$$

Now assume that u is not known in a deterministic sense but may be regarded as a member of a stationary, stochastic process whose correlation sequence $R_u(l)$ is known. Then, upon replacing x by y in (8.14) and substituting (8.21),

$$R_y(l) = \lim_{K\to\infty} \frac{1}{2K+1} \sum_{k=-K}^{K} \sum_{p=-\infty}^{\infty} h(p)u(k-p) \sum_{q=-\infty}^{\infty} h(q)u(k+l-q) \tag{8.22}$$

By interchanging the summations,

$$R_y(l) = \sum_{p=-\infty}^{\infty} h(p) \sum_{q=-\infty}^{\infty} h(q)R_u(l+p-q) \tag{8.23}$$

Equation (8.23) expresses the autocorrelation sequence of the output process in terms of that of the input.

It is frequently also of interest to determine the crosscorrelation between the input and output processes. From (8.15) and (8.21),

$$\begin{aligned} R_{uy}(l) &= \lim_{K\to\infty} \frac{1}{2K+1} \sum_{k=-K}^{K} u(k) \sum_{p=-\infty}^{\infty} h(p)u(k+l-p) \\ &= \sum_{p=-\infty}^{\infty} h(p) \lim_{K\to\infty} \frac{1}{2K+1} \sum_{k=-K}^{K} u(k)u(k+l-p) \\ &= \sum_{p=-\infty}^{\infty} h(p)R_u(l-p) \end{aligned} \tag{8.24}$$

8.4 DISCRETE-PROCESS SPECTRAL DENSITY FUNCTIONS

Just as it was possible to replace the convolution summation (2.6) by the generating-function multiplication (3.15), it is also possible to replace the summations (8.23) and (8.24) by multiplications in a transformed domain. Thus if we let $S_y(z)$ be the generating function of $R_y(l)$ in (8.23), we obtain

$$S_y(z) = \mathcal{G}[R_y(l)] = \sum_{l=-\infty}^{\infty} \sum_{p=-\infty}^{\infty} h(p) \sum_{q=-\infty}^{\infty} h(q)R_u(l + p - q)z^{-l}$$

$$= \sum_{p=-\infty}^{\infty} h(p)z^{p} \sum_{q=-\infty}^{\infty} h(q)z^{-q} \sum_{l=-\infty}^{\infty} R_u(l + p - q)z^{-(l+p-q)}$$

and hence

$$\boxed{S_y(z) = H(z^{-1})H(z)S_u(z)} \tag{8.25}$$

where

$$S_u(z) = \mathcal{G}[R_u(l)] \tag{8.26}$$

The expression $H(z^{-1})$ represents $H(z)$ with z replaced by z^{-1}. Note that if $H(z)$ converges for all $|z| > R_{ch}$, then $H(z^{-1})$ will converge for all $|z| < 1/R_{ch}$. If a region of convergence is to exist for (8.25), it is necessary that

$$1/R_{ch} > |z| > R_{ch}$$

that is, that $R_{ch} < 1$. Since $S_u(z)$ can be regarded as the output of some other system, its region of convergence (if it exists) will similarly be an annulus centered about the point $z = 0$; for example,

$$1/R_{cu} > |z| > R_{cu}$$

where $R_{cu} < 1$. The actual convergence region for (8.25) will then be

$$\min(1/R_{ch}, 1/R_{cu}) > |z| > \max(R_{ch}, R_{cu}) \tag{8.27}$$

Because of its close analogy to the spectral density of continuous-time processes,[3] the function $S_u(z)$ is referred to as the *discrete-process spectral density*.

If we let $S_{uy}(z) = \mathcal{G}[R_{uy}(l)]$, then, from (8.24)

$$S_{uy}(z) = \sum_{l=-\infty}^{\infty} \sum_{p=-\infty}^{\infty} h(p)R_u(l - p)z^{-l}$$

$$= \sum_{p=-\infty}^{\infty} h(p)z^{-p} \sum_{l=-\infty}^{\infty} R_u(l - p)z^{-(l-p)}$$

and hence

$$\boxed{S_{uy}(z) = H(z)S_u(z)} \tag{8.28}$$

for all z such that

$$1/R_{cu} > |z| > \max(R_{ch}, R_{cu}) \tag{8.29}$$

[3] See Y. W. Lee, *op. cit.*, pp. 56–72, or Davenport and Root, *op. cit.*, pp. 89–109.

A number of other system relations are readily obtained in a similar manner. From (8.25) and (8.28),

$$\boxed{S_y(z) = H(z^{-1})S_{uy}(z)} \tag{8.30}$$

From (8.19), (8.24), and (8.18),

$$R_{yu}(l) = \sum_{p=-\infty}^{\infty} h(p)R_u(l + p) \tag{8.31}$$

and hence

$$\boxed{S_{yu}(z) = H(z^{-1})S_u(z)} \tag{8.32}$$

Observe that since the correlation sequence is the inverse generating function of the corresponding spectral density function, it follows from (8.16) that

$$\overline{u^2(k)} = \mathcal{G}^{-1}[S_u(z)]_{l=0} \tag{8.33}$$

Similarly,

$$\overline{u(k)y(k)} = \mathcal{G}^{-1}[S_{uy}(z)]_{l=0} = \mathcal{G}^{-1}[S_{yu}(z)]_{l=0} \tag{8.34}$$

In accordance with (8.26),

$$S_u(z) = \sum_{l=-\infty}^{\infty} R_u(l)z^{-l} \tag{8.35}$$

Hence from (8.18) we obtain

$$\boxed{S_u(z) = S_u(z^{-1})} \tag{8.36}$$

Similarly, from (8.19),

$$\boxed{S_{uy}(z) = S_{yu}(z^{-1})} \tag{8.37}$$

Let us now suppose that we have a system of weighting sequence h, input u, and output y, and we are given the cross-spectral density $S_{ud}(z)$ for the input u and some unspecified time sequence d. It is then a simple matter to determine $S_{yd}(z)$, the cross-spectral density for the system output y, and the time sequence d. From (8.15),

$$R_{yd}(l) = \lim_{K\to\infty} \frac{1}{2K + 1} \sum_{k=-K}^{K} y(k)d(k + l) \tag{8.38}$$

Substituting from (8.21), we have

$$R_{yd}(l) = \lim_{K\to\infty} \frac{1}{2K + 1} \sum_{k=-K}^{K} \sum_{p=-\infty}^{\infty} h(p)u(k - p)d(k + l)$$

Interchanging summations and using (8.38) again yields

$$R_{yd}(l) = \sum_{p=-\infty}^{\infty} h(p)R_{ud}(l + p) \tag{8.39}$$

Then

$$S_{yd}(z) = \sum_{l=-\infty}^{\infty} \sum_{p=-\infty}^{\infty} h(p)R_{ud}(l + p)z^{-l}$$

$$= \sum_{p=-\infty}^{\infty} h(p)z^{p} \sum_{l=-\infty}^{\infty} R_{ud}(l + p)z^{-(l+p)}$$

and hence

$$\boxed{S_{yd}(z) = H(z^{-1})S_{ud}(z)} \tag{8.40}$$

8.5 SYSTEM DESIGN FOR MINIMUM MEAN-SQUARE ERROR

A problem of considerable interest in system theory is that of the design of "optimum" systems, that is, of systems that are "the best" when measured against a specified performance criterion. One approach to the solution of this problem was given by Wiener.[4] It consists of a systematic method for determining the weighting function of a linear system *that is optimum in the sense that the mean-square error between the actual output and the desired output is minimized.* For example, if a system has an input (of known statistical properties) that is corrupted by additive noise,[5] we can determine the system weighting function that will yield an output differing with minimum mean-square error from what the output would be if the input were noise-free. We shall now describe this method for the case of discrete-time systems.[6]

Consider a system described by the weighting sequence $\{h(k)\}$ $\big(h(k) = 0$ for $k > 0\big)$. Let the input be u and the output be y, as shown in Fig. 8-1. Then the difference e between a desired output d and the actual output is represented by

$$e(k) = d(k) - y(k) \tag{8.41}$$

The mean-square error is

$$\overline{e^2(k)} = \overline{d^2(k)} - \overline{d(k)y(k)} - \overline{y(k)d(k)} + \overline{y^2(k)} \tag{8.42}$$

[4] Norbert Wiener, *Extrapolation, Interpolation and Smoothing of Stationary Time Series*, John Wiley and Sons, New York, 1949.

[5] Any unwanted signal is referred to as "noise."

[6] R. H. Barker, "The Theory of Pulse-Monitored Servos and Their Use for Prediction," *Report 1046*, Signals Research and Development Establishment, Christchurch, Hants, England, November 1950.

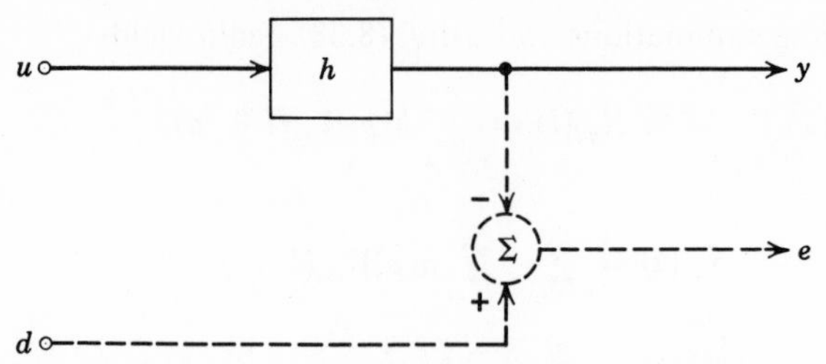

Figure 8-1 Block diagram for design of system with minimum mean-square error.

which, in accordance with (8.33) and (8.34), can be written as

$$\overline{e^2(k)} = \mathscr{G}^{-1}\{S_d(z) - S_{dy}(z) - S_{yd}(z) + S_y(z)\}_{l=0} \tag{8.43}$$

But since

$$y(k) = \sum_{j=-\infty}^{\infty} h(j)u(k-j) \tag{8.44}$$

we have from (8.40)

$$S_{yd}(z) = H(z^{-1})S_{ud}(z) \tag{8.45}$$

and from (8.37)

$$S_{dy}(z) = H(z)S_{du}(z) \tag{8.46}$$

Hence

$$\overline{e^2(k)} = \mathscr{G}^{-1}\{S_d(z) - H(z)S_{du}(z) - H(z^{-1})S_{ud}(z) + H(z)H(z^{-1})S_u(z)\}_{l=0} \tag{8.47}$$

To determine the $H(z)$ that will minimize $\overline{e^2(k)}$ we let $H(z)$ undergo a variation and note the effect on $\overline{e^2(k)}$. When the condition is found from which any variation results in an increase of $\overline{e^2(k)}$, the optimum $H(z)$ will have been determined.

In (8.47), replace $H(z)$ by $H(z) + \varepsilon H_1(z)$ and $\overline{e^2(k)}$ by $\overline{e^2(k)} + \overline{\delta e^2(k)}$. $H_1(z)$ is a nonanticipatory but otherwise arbitrary transfer function and ε is a variational parameter. Upon subtracting (8.47) from the resulting equation we obtain

$$\begin{aligned}\overline{\delta e^2(k)} = \mathscr{G}^{-1}\{&-\varepsilon H_1(z)S_{du}(z) - \varepsilon H_1(z^{-1})S_{ud}(z) + \varepsilon H_1(z)H(z^{-1})S_u(z) \\ &+ \varepsilon H_1(z^{-1})H(z)S_u(z) + \varepsilon^2 H_1(z)H_1(z^{-1})S_u(z)\}_{l=0}\end{aligned} \tag{8.48}$$

If $H(z)$ is to be the desired optimum, it is necessary that

$$\left[\frac{\partial}{\partial\varepsilon}\,\overline{\delta e^2(k)}\right]_{\varepsilon=0} = 0 \tag{8.49}$$

and that the corresponding second derivative be positive, thus indicating that the extremum point at $\varepsilon = 0$ is a minimum (rather than a maximum).

We find

$$\left[\frac{\partial}{\partial \varepsilon}\,\overline{\delta e^2(k)}\right]_{\varepsilon=0} = \mathscr{G}^{-1}\{H_1(z)[H(z^{-1})S_u(z) - S_{du}(z)] + H_1(z^{-1})[H(z)S_u(z) - S_{ud}(z)]\}_{l=0} \quad (8.50)$$

$$\left[\frac{\partial^2}{\partial \varepsilon^2}\,\overline{\delta e^2(k)}\right]_{\varepsilon=0} = \mathscr{G}^{-1}\{2H_1(z)H_1(z^{-1})S_u(z)\}_{l=0} \quad (8.51)$$

The second derivative is positive (and hence we have a minimum point) since the expression on the right in (8.51) may be regarded as the *mean-square output* of a physically realizable system whose weighting sequence is given by h_1 and whose input is represented by u.

If (8.50) is written directly in terms of the correlation and weighting sequences, we obtain, after combining with (8.49),

$$\sum_{k=-\infty}^{\infty} h_1(k)\left[\sum_{p=-\infty}^{\infty} h(p)R_u(l-k+p) - R_{du}(l-k)\right]_{l=0} + \sum_{k=-\infty}^{\infty} h_1(k)\left[\sum_{p=-\infty}^{\infty} h(p)R_u(l+k-p) - R_{ud}(l+k)\right]_{l=0} = 0 \quad (8.52)$$

But since

$$R_u(p-k) = R_u(k-p)$$

and

$$R_{du}(-k) = R_{ud}(k)$$

for $l = 0$, (8.52) reduces simply to

$$2\sum_{k=-\infty}^{\infty} h_1(k)\left[\sum_{p=-\infty}^{\infty} h(p)R_u(k-p) - R_{ud}(k)\right] = 0 \quad (8.53)$$

If $h_1(k)$ is to be the weighting sequence of a nonanticipatory system, it must vanish for all $k < 0$. Hence (8.53) is clearly satisfied for all $k < 0$. However, except for the requirement of nonanticipation, $h_1(k)$ may be completely arbitrary. This implies that to satisfy (8.53) for $k \geq 0$, it is *necessary* that

$$\sum_{p=-\infty}^{\infty} h(p)R_u(k-p) - R_{ud}(k) = 0 \quad \text{for } k \geq 0 \quad (8.54)$$

Now suppose that we form the generating function of the expression on the left in (8.54). Since this expression is nonzero for $k < 0$ and is zero for $k \geq 0$, its generating function will converge for all z *inside* the circle $|z| = 1$.[7] This implies that the generating function of (8.54), that is,

$$H(z)S_u(z) - S_{ud}(z)$$

may have no poles inside the circle defined by $|z| = 1$, (i.e., the so-called *unit circle*).

[7] Cf. Section 3.2.

Let

$$S_u(z) = S_u^+(z)S_u^-(z) \tag{8.55}$$

where $S_u^+(z)$ contains all the poles and zeros *inside* the unit circle, and $S_u^-(z)$ contains all those which lie outside. Then the generating function of the left side of (8.54) can be written as

$$S_u^-(z)\left[H(z)S_u^+(z) - \frac{S_{ud}(z)}{S_u^-(z)}\right]$$

It follows from the foregoing that the expression in the brackets may have no poles inside the unit circle. However, $H(z)S_u^+(z)$ can have poles *only* inside the unit circle.[8] Therefore, the terms in the partial-fraction expansion of $S_{ud}(z)/S_u^-(z)$ due to poles inside the unit circle must *cancel* the terms in the partial-fraction expansion of $H(z)S_u^+(z)$. We may write

$$\frac{S_{ud}(z)}{S_u^-(z)} = \left[\frac{S_{ud}(z)}{S_u^-(z)}\right]_+ + \left[\frac{S_{ud}(z)}{S_u^-(z)}\right]_- \tag{8.56}$$

where the subscript "+" denotes the collection of terms in the partial-fraction expansion associated with poles *inside* the unit circle, and the subscript "−" refers to the terms associated with poles *outside* the unit circle. Then

$$H(z)S_u^+(z) - \left[\frac{S_{ud}(z)}{S_u^-(z)}\right]_+ = 0 \tag{8.57}$$

Solving for $H(z)$, we obtain

$$\boxed{H(z) = \frac{1}{S_u^+(z)}\left[\frac{S_{ud}(z)}{S_u^-(z)}\right]_+} \tag{8.58}$$

Equation (8.58) gives the transfer function of the nonanticipatory, linear, discrete-time system that yields the minimum mean-square error between its output y and some desired output d. The transfer function is completely specified if the spectral densities $S_u(z)$ and $S_{ud}(z)$, or the corresponding correlation sequences, are given.

Let us now consider the case where the input u consists of a signal f plus additive noise n of zero mean and the desired output is the signal f. We have

$$u(k) = f(k) + n(k) \tag{8.59}$$

$$R_u(l) = \overline{[f(k) + n(k)][f(k+l) + n(k+l)]} \tag{8.60}$$

$$= R_f(l) + R_{fn}(l) + R_{nf}(l) + R_n(l) \tag{8.61}$$

[8] $H(z)$ must have all its poles inside the unit circle since it is to be a nonanticipatory system whose weighting sequence $\{h(k)\}$ converges to zero for large k. The poles of $S_u^+(z)$ are inside the unit circle by construction.

and

$$S_u(z) = S_f(z) + S_{fn}(z) + S_{nf}(z) + S_n(z) \tag{8.62}$$

Also

$$R_{ud}(l) = \overline{[f(k) + n(k)]f(k + l)} \tag{8.63}$$

$$= R_f(l) + R_{nf}(l) \tag{8.64}$$

and

$$S_{ud}(z) = S_f(z) + S_{nf}(z) \tag{8.65}$$

Upon substituting (8.62) and (8.65) in (8.58), we obtain the transfer function of the linear system that is *optimum* in the sense of minimizing the mean-square error due to the noise.[9]

To determine the value of the mean-square error when the system has the optimum transfer function, (8.58), we substitute (8.58) in (8.47). After simplification we obtain

$$\boxed{\overline{e^2(k)}_{\text{opt}} = \mathscr{G}^{-1}[S_d(z) - H(z)_{\text{opt}} S_{du}(z)]_{l=0}} \tag{8.66}$$

Example

An optimum linear system is sought that has a desired output f when the input consists of a signal f plus additive, zero-mean noise n. The signal process f is characterized by the autocorrelation sequence $R_f(l) = (0.8)^{|l|}$. The noise has a spectral density $S_n(z) = 0.49$. Signal and noise are uncorrelated.

We write

$$S_f(z) = \mathscr{G}[(0.8)^{|l|}]$$

$$= \frac{1}{1 - 0.8z^{-1}} + \frac{0.8z}{1 - 0.8z}$$

$$= \frac{-0.45z}{(z - 0.8)(z - 1.25)}$$

$$S_{nf}(z) = 0$$

$$S_u(z) = \frac{-0.45z}{(z - 0.8)(z - 1.25)} + 0.49$$

$$= \frac{(0.7)^2(z - 0.3875)(z - 2.5805)}{(z - 0.8)(z - 1.25)}$$

Hence

$$S_u^+(z) = \frac{0.7(z - 0.3875)}{z - 0.8}$$

$$S_u^-(z) = \frac{0.7(z - 2.5805)}{z - 1.25}$$

[9] For obvious reasons, the resulting system is frequently also referred to as the *optimum linear filter*.

Since $S_{nf}(z) = 0$,

$$S_{ud}(z) = S_f(z)$$

Substituting in (8.58), we have

$$H(z) = \frac{z - 0.8}{(0.7)(z - 0.3875)}\left[\frac{-0.45z}{0.7(z - 2.5805)(z - 0.8)}\right]_+$$

We next perform the partial-fraction expansion of the expression in the brackets and retain only the term corresponding to the pole inside the circle $|z| = 1$. After simplification this then yields the desired transfer function

$$H(z) = \frac{0.413}{z - 0.3875}, \qquad |z| > 0.3875$$

The corresponding weighting sequence is given by

$$\begin{aligned} h(k) &= 0.413(0.3875)^{k-1} && \text{for } k \geq 1 \\ &= 0 && \text{for } k \leq 0 \end{aligned}$$

The minimum mean-square error is then found by (8.66)

$$\begin{aligned} \overline{e^2(k)}_{\text{opt}} &= \mathscr{G}^{-1}[S_f(z) - H(z)S_f(z)]_{l=0} \\ &= \mathscr{G}^{-1}\left[\frac{-0.45z}{(z - 1.25)(z - 0.3875)}\right]_{l=0} \\ &= 0.521 \end{aligned}$$

8.6 STOCHASTIC FINITE-STATE SYSTEMS

All the systems that we have considered thus far have been *deterministic* in nature; that is, the systems possessed the property that their states at any time t could be predicted *with certainty* from knowledge of their states at some prior time t_0 and knowledge of their inputs during the interval $[t_0, t)$. In those cases where the input was given only in statistical terms, the description of state or output could, of course, be obtained also only in statistical terms. The characterization of the system itself, however, was deterministic and could be given either in the form of state equations such as (1.4) and (1.5) or (if linear and stationary) as transfer functions such as (3.17). Deterministic systems with stochastic inputs were considered in the preceding sections of this chapter.

We shall now consider a class of *stochastic* system, that is, systems whose state and output can be described only statistically, even if the input is deterministic. For such systems there are no state equations such as

(1.5). Instead we characterize them by giving the *probabilities* that the system will be in any specified state at time t, given the state at time $t_0 < t$. Characterization in terms of these *state transition probabilities* is simple in form provided we restrict consideration to systems that can assume only a finite number of discrete states, that is, to so-called *finite-state systems.*[10] This is a severe restriction with regard to systems in general; however, it permits us to study a class of interesting systems by means of the techniques developed in the preceding chapters.

By considering only finite-state systems, the customary vector notation for a state may be replaced by a simple scalar numbering; that is, we can denote the states of the system by $i = 1, 2, \ldots, \nu$, where ν (finite) is the total number of states. Since the states are discrete, the transitions from state to state must occur at discrete instants of time. Without loss of generality we can, therefore, take time as a discrete variable and write $i(k)$ to indicate that the system is in state i at time k, $k = 0, 1, 2, \ldots, \infty$.

The probability that the system is in state i at time k will be referred to as the *state probability* and denoted by $p_i(k)$, $i = 1, 2, \ldots, \nu$. A stochastic, finite-state system is characterized by its *transition probabilities* $q_{ji}(k)$ that relate the state probabilities of the system at time $k + 1$ to the state probabilities at time k. We shall consider only systems for which these transition probabilities are independent of the past history of the system; that is, if at time k the system is in state i, there is a probability $q_{ji}(k)$ that it will next be in state j, *irrespective of how it arrived at state i.* Systems that possess this property will be referred to as *Markov systems*, or, in conformance with the mathematical literature, as *Markov chains.*[11]

We can write a simple recursion equation for the state probabilities of a Markov system. Consider the probability $p_j(k + 1)$ that the system be in a particular state j at time $k + 1$. Now at time k the system may be in any state i with the probability $p_i(k)$, $i = 1, 2, \ldots, \nu$. For each state i there is a transition probability q_{ji} that the system will make the transition from this state to state j. Hence the probability that the system will be in a particular state j at time $k + 1$ is equal to the sum of the products $q_{ji}(k)p_i(k)$; that is,

$$p_j(k + 1) = \sum_{i=1}^{\nu} q_{ji}(k)p_i(k), \qquad j = 1, 2, \ldots, \nu \tag{8.67}$$

[10] The systems considered earlier were restricted only to a *finite-dimensional* state space of dimension n. There was no bound on the number of admissible states within this space.

[11] See, for example, J. G. Kemeny and J. L. Snell, *Finite Markov Chains*, D. Van Nostrand Company, Princeton, N.J., 1960; or L. Takacs, *Stochastic Processes*, Methuen and Co., Ltd., London, 1960.

In vector form,

$$\mathbf{p}(k+1) = \mathbf{Q}(k)\mathbf{p}(k) \tag{8.68}$$

where $\mathbf{p}$ is a ν-dimensional state probability vector and $\mathbf{Q}$ is a $\nu \times \nu$ *transition probability matrix*. If all the q_{ji} are independent of k, we have

$$\mathbf{p}(k+1) = \mathbf{Q}\mathbf{p}(k) \tag{8.69}$$

and the Markov system is said to be *stationary*.

The *state probability equations* (8.68) and (8.69) are, respectively, of precisely the same form as the nonstationary and stationary state equations for linear, deterministic, input-free systems. The systems described by

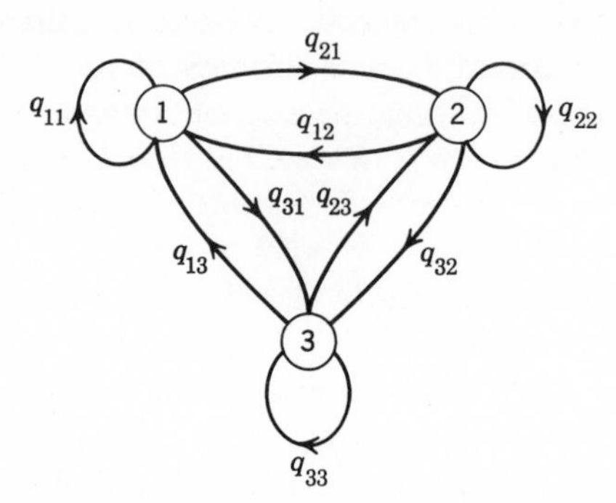

Figure 8-2 A three-state Markov system.

(8.68) and (8.69), however, need not be linear; they must merely be finite-state Markov systems. We thus have the remarkable fact that a class of nonlinear systems characterized in terms of state transition probabilities is governed by linear equations.

Markov systems can be represented graphically by so-called state diagrams. These are signal flow graphs in which the nodes represent the states i and the directed branches represent the transition probabilities q_{ji}, $i = 1, 2, \ldots, \nu$. The state diagram of a stationary three-state Markov system is shown in Fig. 8-2.

It is clear from the nature of transition probabilities that the elements of the matrix $\mathbf{Q}$ must satisfy the following conditions, whether the matrix is stationary or nonstationary:[12]

$$q_{ji} \geq 0 \quad \text{for all } i, j \tag{8.70}$$

and

$$\sum_{j=1}^{\nu} q_{ji} = 1 \quad \text{for all } i \tag{8.71}$$

[12] A matrix satisfying (8.70) and (8.71) is known as a *stochastic matrix*. Any stochastic matrix may serve as a transition probability matrix.

The solutions of (8.68) and (8.69) are easily obtained by following the method of Section 2.4. We find

$$\mathbf{p}(k) = \mathbf{Q}(k-1)\mathbf{Q}(k-2)\ldots\mathbf{Q}(0)\mathbf{p}(0) \tag{8.72}$$

if $\mathbf{Q}$ is nonstationary, and

$$\mathbf{p}(k) = \mathbf{Q}^k\mathbf{p}(0) \tag{8.73}$$

if $\mathbf{Q}$ is stationary. Equations (8.72) and (8.73) show that the state probabilities of a Markov system are completely determined for all $k \geq 0$ if we know the transition probability matrix $\mathbf{Q}$ and the initial state probability vector $\mathbf{p}(0)$.

The evaluation of (8.73) for large k is best carried out by means of the generating function technique described in Section 3.5.[13] If we let $\mathbf{P}(z)$ denote the vector generating function of $\mathbf{p}(k)$, we have from (8.69)

$$z\mathbf{P}(z) - z\mathbf{p}(0) = \mathbf{Q}\mathbf{P}(z) \tag{8.74}$$

and hence

$$\mathbf{P}(z) = [\mathbf{I} - \mathbf{Q}z^{-1}]^{-1}\mathbf{p}(0) \tag{8.75}$$

Since the components of $\mathbf{p}(k)$ may not exceed unity, the generating function $\mathbf{P}(z)$ converges for all $|z| > |$.

Equation (8.75) readily permits the determination of the probability that the system be in a state i at any $k > 0$, given the initial distribution $\mathbf{p}(0)$.

Example

It is desired to determine the probability that the system of Fig. 8-3 be in state 3 at $k = 10$, given that the system is in state 1 at $k = 0$. The transition matrix is as follows:

$$\mathbf{Q} = \begin{bmatrix} \frac{1}{2} & \frac{1}{3} & 0 \\ \frac{1}{2} & 0 & \frac{1}{5} \\ 0 & \frac{2}{3} & \frac{4}{5} \end{bmatrix}$$

The initial state probability vector is given by

$$\mathbf{p}(0) = \begin{bmatrix} 1 \\ 0 \\ 0 \end{bmatrix}$$

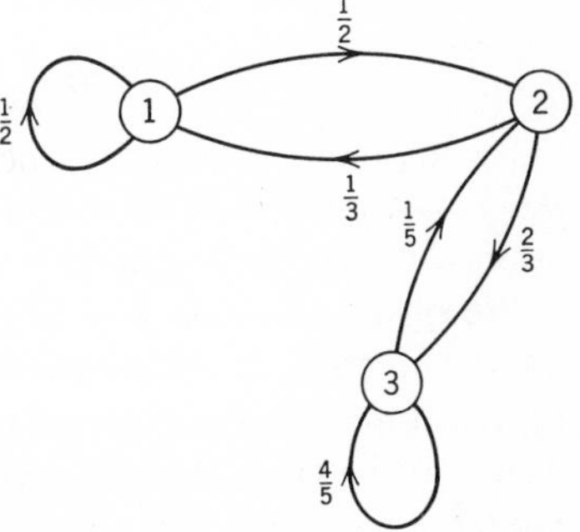

Figure 8-3 Transition probability diagram for example.

[13] R. W. Sittler, "System Analysis of Discrete Markov Processes," *IRE Trans. Circuit Theory*, vol. CT-3, no. 1, pp. 257–266, December 1956.

Using (8.75) we write

$$\mathbf{P}(z) = \begin{bmatrix} 1 - \frac{1}{2}z^{-1} & -\frac{1}{3}z^{-1} & 0 \\ -\frac{1}{2}z^{-1} & 1 & -\frac{1}{5}z^{-1} \\ 0 & -\frac{2}{3}z^{-1} & 1 - \frac{4}{5}z^{-1} \end{bmatrix}^{-1} \begin{bmatrix} 1 \\ 0 \\ 0 \end{bmatrix}$$

$$= \frac{z^{-2}}{\Delta} \begin{bmatrix} z^2 - \frac{4}{5}z - \frac{2}{15} & \frac{1}{3}z - \frac{4}{15} & \frac{1}{15} \\ \frac{1}{2}z - \frac{2}{5} & z^2 - \frac{13}{10}z + \frac{2}{5} & \frac{1}{5}z - \frac{1}{10} \\ \frac{1}{3} & \frac{2}{3}z - \frac{1}{3} & z^2 - \frac{1}{2}z - \frac{1}{6} \end{bmatrix} \begin{bmatrix} 1 \\ 0 \\ 0 \end{bmatrix}$$

where

$$\Delta = z^{-3}(z - 1)(z - 0.6217)(z + 0.3217)$$

Hence

$$\mathbf{P}(z) = \frac{z}{(z - 1)(z - 0.6217)(z + 0.3217)} \begin{bmatrix} z^2 - \frac{4}{5}z - \frac{2}{15} \\ \frac{1}{2}z - \frac{2}{5} \\ \frac{1}{3} \end{bmatrix}$$

If we desire only the probability of being in state 3, we select the third component of this column vector; that is,

$$p_3(z) = \frac{\frac{1}{3}z}{(z - 1)(z - 0.6217)(z + 0.3217)}$$

Taking the inverse generating function then yields

$$p_3(k) = 0.6667 - 0.934(0.6217)^k + 0.2673(-0.3217)^k \quad \text{for } k \geq 0$$

At $k = 10$,

$$\begin{aligned} p_3(10) &= 0.6667 - 0.934(0.0087) + 0.2673(0.0000) \\ &= 0.6586 \end{aligned}$$

Evaluation of $p_3(k)$ for a few other values of k shows that $p_3(0) = p_3(1) = 0$, $p_3(2) = 0.333$, and $p_3(3) = 0.433$. These values can be readily verified by a step-by-step examination of the states in Fig. 8-3.

We observe that at $k = 10$, $p_3(k)$ has nearly reached its "steady state" value of 0.6667. In other words, for $k \gg 10$, the system is in state 3 with a probability of 0.6667 and in some other state with a probability of 0.3333. The knowledge of the initial state (i.e., state 1 at $k = 0$) fades into insignificance as $k \to \infty$, as we might expect.

The values of the state probabilities for large k can, of course, be obtained directly by applying the final-value theorem (3.68) to the components of $\mathbf{P}(z)$. We find $p_1(\infty) = 0.1333$, $p_2(\infty) = 0.2000$, and $p_3(\infty) = 0.6667$. These are the *limiting state probabilities* of the system.

8.7 TYPES OF MARKOV SYSTEMS

Let us group the states of a Markov system into *state sets* such that within each state set the system can change from every state to every other state, directly or indirectly. The state sets will be mutually exclusive; that is, each state will be a member of one and only one such set. We can then introduce the following definitions:

1. A state set that, once entered, can never be left is called an *ergodic set*. Its states are referred to as *recurrent states*.

2. A state set that, once left, can never again be entered is called a *transient set*. All the states that do not belong to ergodic sets belong to transient sets and are called *transient states*. The limiting state probabilities of all transient states are zero.

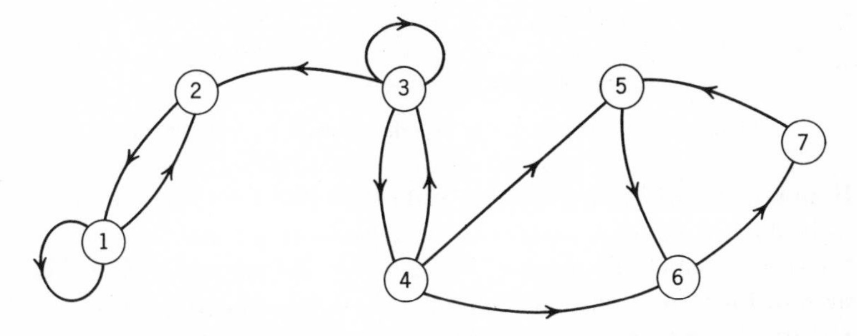

Figure 8-4 A seven-state Markov system. (The transition probabilities of all branches shown are presumed to be nonzero.)

A Markov system can have many state sets, both transient and ergodic; as a minimum it must have at least one ergodic set. A system that has only one ergodic set is called an *ergodic system*. If a system has more than one ergodic set, but no transient sets, the state sets are disjoint and the system can be separated into a number of independent ergodic subsystems, one for each ergodic set. In Fig. 8-4, there are two ergodic sets, one consisting of states 1 and 2 and the other of states 5, 6, and 7. States 3 and 4 form a transient set. The system of Fig. 8-5 has only two ergodic sets and can be separated into two disjoint systems.

For an ergodic system, all the limiting state probabilities are independent of the system's initial state. This is, however, not true for a nonergodic system. If the probability that a nonergodic system may enter a particular ergodic set is nonzero, all the limiting state probabilities of the states in this set will be nonzero.[14] Thus if the system of Fig. 8-4 is started in state 2, the limiting state probabilities of states 1 and 2 will be nonzero while those of states 5, 6, and 7 will all be zero. However, if the system starts in either state 3 or 4, the limiting state probabilities of the states in *both* ergodic sets will be nonzero. (In either case, the sum of the limiting state probabilities must, of course, be equal to unity.)

Occasionally an ergodic set will consist of only a single state; that is, there is a state that once entered can never be left. Such a state is called an *absorbing state*.

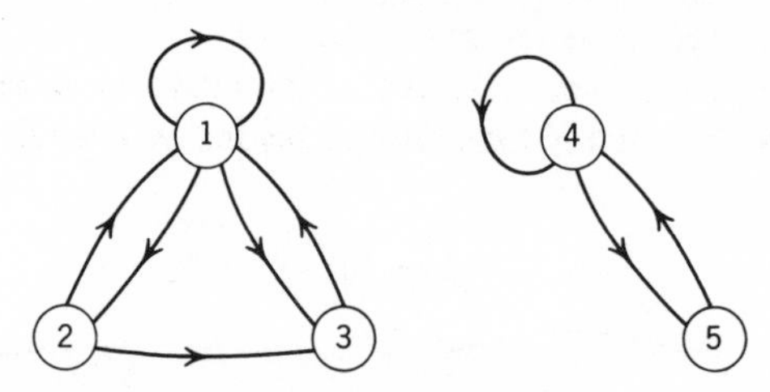

Figure 8-5 System with two ergodic sets and no transient sets.

If an ergodic set has at least one state such that the system *must* return to this state at periodic intervals, the ergodic set is said to be *periodic*; otherwise it is said to be *regular*. For example, the ergodic set 5, 6, 7 in the system of Fig. 8-4 is periodic; the set 1, 2, however, is regular.

Let us now return to the generating function of the state probability vector given by (8.75). The poles of (8.75) are the roots of the characteristic equation

$$|\mathbf{I} - \mathbf{Q}z^{-1}| = 0 \tag{8.76}$$

These are, of course, identical with the eigenvalues of the transition probability matrix $\mathbf{Q}$. If we invert (8.75), we obtain the components of the state probability vector $p_i(k)$ as functions of k. Since $|p_i(k)| \leq 1$ for all i and k, it follows that none of the roots of (8.76) (i.e., the eigenvalues of $\mathbf{Q}$) may have a magnitude greater than unity. Also since all roots of magnitude less than unity contribute transient terms, which vanish as $k \to \infty$, every Markov system must have at least one root that is equal to unity. More precisely, *the characteristic equation* (8.76) *of a Markov system will have as many roots equal to unity as the system has ergodic sets.*

If the characteristic equation of a Markov system contains a factor of the form $(z^\alpha - 1)$, where α is an integer greater than one, then the system

[14] An exception occurs in the case of states with infinite mean recurrence times, that is, so-called *null states*. These states, though recurrent, have limiting state probabilities that are zero. See L. Takacz, *op. cit.*

contains an ergodic set that is periodic with period α. There will be one such factor for each periodic set.

Example

The system of Fig. 8-6 is in state 1 at $k = 0$. We wish to find $p_1(k)$, the probability that the system is in state 1 for any $k \geq 0$. We have

$$\mathbf{Q} = \begin{bmatrix} \frac{1}{2} & \frac{2}{3} & 0 \\ \frac{1}{2} & 0 & 0 \\ 0 & \frac{1}{3} & 1 \end{bmatrix} \qquad \mathbf{p}(0) = \begin{bmatrix} 1 \\ 0 \\ 0 \end{bmatrix}$$

and

$$[\mathbf{I} - \mathbf{Q}z^{-1}]^{-1} = \frac{1}{\Delta}\begin{bmatrix} z(z-1) & \dfrac{2(z-1)}{3} & 0 \\ \dfrac{z-1}{2} & (z-\frac{1}{2})(z-1) & 0 \\ \dfrac{1}{6} & \dfrac{z-\frac{1}{2}}{3} & z^2 - \dfrac{z}{2} - \dfrac{1}{3} \end{bmatrix}$$

where

$$\Delta = z^{-1}(z-1)(z^2 - \tfrac{1}{2}z - \tfrac{1}{3})$$

The characteristic equation will have only one root of unit magnitude, namely $z = 1$; hence the system has only a single, regular, ergodic set. We obtain

$$\mathbf{P}(z) = \begin{bmatrix} \dfrac{z^2}{z^2 - \frac{1}{2}z - \frac{1}{3}} \\ \dfrac{\frac{1}{2}z}{z^2 - \frac{1}{2}z - \frac{1}{3}} \\ \dfrac{\frac{1}{6}z}{(z-1)(z^2 - \frac{1}{2}z - \frac{1}{3})} \end{bmatrix}$$

The inversion of $P_1(z)$ then yields

$$p_1(k) = 0.699(0.879)^k + 0.301(-0.379)^k, \qquad k \geq 0$$

We find

$$p_1(0) = 1$$
$$p_1(1) = 0.5$$
$$p_1(2) = 0.583, \quad \text{etc.}$$

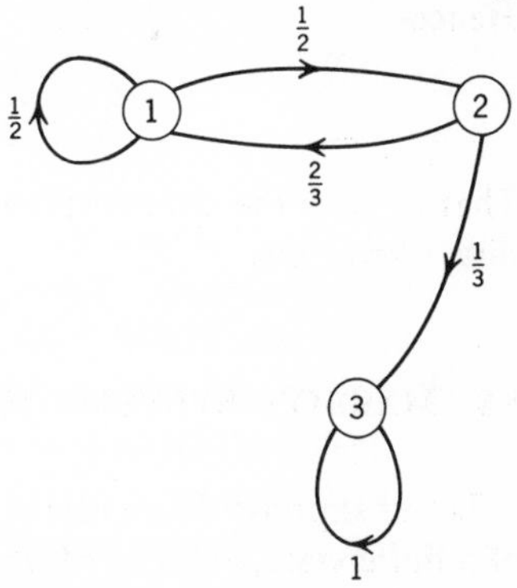

Figure 8-6 Transition probability diagram for system with an absorbing state.

In contrast with the example in Section 8.6, here $p_1(k)$ approaches zero for large k. This is, of course, to be expected since from Fig. 8-6 it is clear that states 1 and 2 are transient states and state 3 is an absorbing state.

Once the system enters state 3, no further state transitions will occur. Let us calculate the mean time to quiescence for this system, that is, the time until all transitions stop. Let $p_e(k)$ be the probability of *initial entry* into state 3 at time k. Then the mean time of first entering (which, for an absorbing state, is the same as the mean time to quiescence) is in accordance with (8.7) given by

$$\bar{k} = \sum_{k=0}^{\infty} k p_e(k)$$

But $p_e(k)$ is simply the difference[15] between $p_3(k)$ and $p_3(k-1)$, that is,

$$p_e(k) = p_3(k) - p_3(k-1) = \nabla p_3(k)$$

Hence if we let

$$\mathscr{G}[p_e(k)] = P_e(z)$$

then from (3.45)

$$\begin{aligned} P_e(z) &= (1 - z^{-1})P_3(z) \\ &= \frac{\frac{1}{6}}{z^2 - \frac{1}{2}z - \frac{1}{3}} \end{aligned}$$

Now from (3.31),

$$\begin{aligned} \sum_{k=0}^{\infty} k p_e(k) &= \sum_{k=0}^{\infty} k p_e(k) z^{-k} \bigg|_{z=1} \\ &= -z \frac{\partial}{\partial z} P_e(z) \bigg|_{z=1} \end{aligned}$$

Hence

$$\bar{k} = \frac{z(2z - \frac{1}{2})/6}{(z^2 - \frac{1}{2}z - \frac{1}{3})^2} \bigg|_{z=1} = 9$$

That is, with the system starting in state 1, the mean time to quiescence is nine transitions.

8.8 MARKOV SYSTEMS WITH COST FUNCTIONS

Let us suppose that a *cost* c_i is associated with each state i, $i = 1, 2, \ldots, \nu$ of a finite-state, ergodic Markov system. As the system undergoes transitions from state to state, a system *operating cost*, is accumulated due to the costs[16] associated with these states. If a system is initially in state i, we can, of course, not predict the precise operating cost to some future time k.

[15] For an absorbing state i, the probability $p_i(k)$ of being in state i at time k is the *cumulative probability* of having entered the state at any time equal to or less than k.

[16] The "costs" c_i may be positive or negative. A negative cost will, of course, represent a "profit."

However, if we know the transition probabilities, we can determine the *mean cost increase* over a specified number of state transitions for a system initially in state i.

Let us denote the mean cost increase of the system over k transitions, starting at state i, by $f_i(k)$, $i = 1, 2, \ldots, \nu$. Now suppose that for a particular value of k, all the $f_i(k)$ are known and that it is desired to determine $f_i(k+1)$. The quantity $f_i(k+1)$ represents the cost increase for the system after $k+1$ transitions, starting from state i. We shall adopt the convention that the cost associated with a state is "paid" just as the system *leaves* this state, that is, during the transition to the next state.[17] Hence for all i, $f_i(0) = 0$ and $f_i(1) = c_i$.

Consider a system, currently in state i, whose mean cost increase over the next $k+1$ transitions is to be determined. As the system leaves state i, it accumulates the cost increment c_i. After the first transition, with only k transitions now remaining, the system may be in any state j, $j = 1, 2, \ldots, \nu$ with probability q_{ji}. The mean cost increase for the remaining k transitions from any state j is then given by $f_j(k)$. Hence the mean cost increase for starting a system in state i and continuing for a time of $k+1$ transitions is equal to the cost c_i plus the weighted sum of the mean costs that would be incurred if the process were started in any state j and continued for a time of only k transitions; that is,

$$f_i(k+1) = c_i + \sum_{j=1}^{\nu} q_{ji} f_j(k), \qquad \begin{array}{l} i = 1, 2, \ldots, \nu \\ k = 0, 1, \ldots, \infty \end{array} \tag{8.77}$$

In matrix notation,

$$\mathbf{f}(k+1) = \mathbf{Q}'\mathbf{f}(k) + \mathbf{c} \tag{8.78}$$

where $\mathbf{f}(k)$ and $\mathbf{c}$ are ν-dimensional column vectors, and $\mathbf{Q}'$ is the transpose of the transition matrix $\mathbf{Q}$.

Let $\mathbf{F}$ denote the generating function of $\mathbf{f}$. Since $\mathbf{c}$ is a constant vector, its generating function is given simply by $[z/(z-1)]\mathbf{c}$. Then from (3.41),

$$z\mathbf{F}(z) - z\mathbf{f}(0) = \mathbf{Q}'\mathbf{F}(z) + \frac{z}{z-1}\mathbf{c} \tag{8.79}$$

$$[z\mathbf{I} - \mathbf{Q}']\mathbf{F}(z) = z\mathbf{f}(0) + \frac{z}{z-1}\mathbf{c}$$

$$\mathbf{F}(z) = z[z\mathbf{I} - \mathbf{Q}']^{-1}\mathbf{f}(0) + \frac{z}{z-1}[z\mathbf{I} - \mathbf{Q}']^{-1}\mathbf{c} \tag{8.80}$$

where $\mathbf{f}(0)$ is the initial value of $\mathbf{f}$. Since $\mathbf{f}(k)$ is the *increase* in cost over an

[17] The cost is "paid" even if after the time of transition the system returns to the same state in which it resided just prior to the transition.

interval of k transitions, $\mathbf{f}(0) = 0$. Thus

$$\mathbf{F}(z) = \frac{z}{z-1}[z\mathbf{I} - \mathbf{Q}']^{-1}\mathbf{c} \tag{8.81}$$

The inverse generating function of (8.81) then yields the components $f_i(k)$, the cost increases for a process starting in state i, $i = 1, 2, \ldots, \nu$, and continuing for k transitions.

Example

Consider the two-state Markov process characterized by the transition matrix

$$\mathbf{Q} = \begin{bmatrix} \frac{1}{2} & \frac{1}{3} \\ \frac{1}{2} & \frac{2}{3} \end{bmatrix}$$

Let the state cost vector be given by

$$\mathbf{c} = \begin{bmatrix} 2 \\ 3 \end{bmatrix}$$

From (8.81),

$$\mathbf{F}(z) = \frac{z}{z-1}\begin{bmatrix} z - \frac{1}{2} & -\frac{1}{2} \\ -\frac{1}{3} & z - \frac{2}{3} \end{bmatrix}^{-1}\begin{bmatrix} 2 \\ 3 \end{bmatrix}$$

$$= \frac{z}{z-1}\begin{bmatrix} \dfrac{z - \frac{2}{3}}{(z-1)(z-\frac{1}{6})} & \dfrac{\frac{1}{2}}{(z-1)(z-\frac{1}{6})} \\ \dfrac{\frac{1}{3}}{(z-1)(z-\frac{1}{6})} & \dfrac{z - \frac{1}{2}}{(z-1)(z-\frac{1}{6})} \end{bmatrix}\begin{bmatrix} 2 \\ 3 \end{bmatrix}$$

$$= \begin{bmatrix} \dfrac{2z^2 + z/6}{(z-1)^2(z-\frac{1}{6})} \\ \dfrac{3z^2 - 5z/6}{(z-1)^2(z-\frac{1}{6})} \end{bmatrix}$$

Upon inversion,

$$f_1(k) = \tfrac{13}{5}k - \tfrac{18}{25} + \tfrac{18}{25}(\tfrac{1}{6})^k$$

$$f_2(k) = \tfrac{13}{5}k + \tfrac{12}{25} - \tfrac{12}{25}(\tfrac{1}{6})^k \quad \text{for } k \geq 0$$

As k becomes large, the cost increase per transition approaches the same value, $\frac{13}{5}$, irrespective of whether we started in state 1 or in state 2. This is, of course, to be expected since for an ergodic Markov system, the state probabilities approach limits that are independent of the initial state as $k \to \infty$. The mean cost increase per transition then becomes simply equal to the sum of the state costs weighted by the limiting state probabilities.

Note, however, that as $k \to \infty$, the total cost incurred by starting in state 1 will always be less by $\frac{6}{5}$ than if the system were started in state 2.

8.9 CONTROLLABLE MARKOV SYSTEMS

Thus far we have devoted all our attention to input-free Markov systems —that is, systems whose state-to-state transitions were governed entirely by a single, fixed, transition probability matrix. The transitions took place automatically and were not subject to external control. In this section we shall now consider Markov systems with inputs. Two problems are of particular interest: (1) the behavior of a Markov system when a stochastic input is applied, and (2) the determination of the input required to achieve optimum control of a system, given a specified performance criterion. The second of these will be taken up in the next section.

Our systems will still be finite-state systems. For simplicity, we shall also restrict the inputs to a finite set. The inputs can then be denoted simply by u_l, $l = 1, 2, \ldots, \alpha$, where α is the total number of permissible inputs. We shall stipulate that an input must be applied prior to every system transition. Each applied input will determine the state cost incurred upon leaving the present state as well as the transition probabilities that govern the transition to the next state. Thus for a set of α possible distinct inputs, the system will have α state-cost vectors, $\mathbf{c}^l$, and α transition-probability matrices, $\mathbf{Q}_l$. The cost vectors and transition probabilities need, of course, not all be different.

Let us consider an ergodic Markov system that can be controlled by an input. A total of α input values are permissible. We let the input to the system be a stochastic sequence $\{u(k)\}$, consisting of the input values u_l, $l = 1, 2, \ldots, \alpha$. The values u_l are presumed to occur in accordance with a given, discrete, first-order probability density $\pi(l)$ and to be statistically independent.

It is a simple matter to modify the transition equation (8.69) to take into account the probabilistic nature of the transition-probability matrix. We write

$$\mathbf{p}(k + 1) = \sum_{l=1}^{\alpha} \mathbf{Q}_l \pi(l) \mathbf{p}(k) \tag{8.82}$$

$$= \mathbf{\Omega} \mathbf{p}(k) \tag{8.83}$$

where

$$\mathbf{\Omega} = \sum_{l=1}^{\alpha} \mathbf{Q}_l \pi(l) \tag{8.84}$$

is the *mean transition-probability matrix* with which the system is operating when controlled by the stochastic input sequence $\{u(k)\}$.

For the cost-transition equation corresponding to (8.78) we write

$$\mathbf{f}(k+1) = \sum_{l=1}^{\alpha} \pi(l)[\mathbf{Q}_l'\mathbf{f}(k) + \mathbf{c}^l] \tag{8.85}$$

$$= \mathbf{\Omega}'\mathbf{f}(k) + \boldsymbol{\sigma} \tag{8.86}$$

where

$$\boldsymbol{\sigma} = \sum_{l=1}^{\alpha} \pi(l)\mathbf{c}^l \tag{8.87}$$

is now the *mean cost vector* for the system.

With the cost-transition equation in the form (8.86), we can write, using the result of (8.78)–(8.81),

$$\mathbf{F}(z) = \frac{z}{z-1}[z\mathbf{I} - \mathbf{\Omega}']^{-1}\boldsymbol{\sigma} \tag{8.88}$$

From this last equation we can determine the cost increase for the system starting in any given state and operating for k transitions under the control of statistically independent input values of given discrete probability density.

8.10 OPTIMUM CONTROL OF MARKOV SYSTEMS

We next consider the case where a sequence of input values is to be applied to a controllable finite-state Markov system so as to optimize its performance. We shall use the cost function described earlier as our performance measure. Optimum control will be achieved when the cost function is minimized.

With the system in any state i, $i = 1, 2, \ldots, \nu$, and k transitions remaining, we shall determine the input value u_l that must be applied so that the cost $f_i(k)$ will be a minimum. We shall refer to the selected values of u_l for the states i at any time k as our *control decision*, $\mathbf{d}(k)$, where $\mathbf{d}(k)$ is a ν-dimensional vector. A sequence of control decisions will be called a *control policy*.

We shall now introduce the method of optimization by means of an illustrative example.

Example

Consider a two-state ergodic system that can be controlled by means of two possible input values, u_1 and u_2. If u_1 is applied, we have

$$\mathbf{Q}_1 = \begin{bmatrix} \frac{1}{2} & \frac{1}{3} \\ \frac{1}{2} & \frac{2}{3} \end{bmatrix}, \qquad \mathbf{c}^1 = \begin{bmatrix} 6 \\ 18 \end{bmatrix} \tag{8.89}$$

If u_2 is applied,

$$\mathbf{Q}_2 = \begin{bmatrix} \frac{3}{4} & \frac{1}{2} \\ \frac{1}{4} & \frac{1}{2} \end{bmatrix}, \qquad \mathbf{c}^2 = \begin{bmatrix} 7 \\ 21 \end{bmatrix} \tag{8.90}$$

A state diagram of the system is shown in Fig. 8-7. The transition probabilities and state costs marked with a superscript 1 apply when the input is u_1; those marked with a 2 apply when the input is u_2.

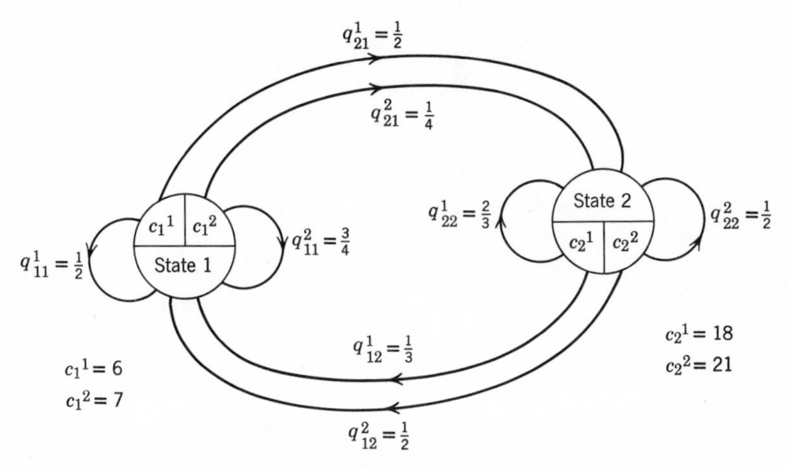

Figure 8-7 Illustrative example. Two-control, two-state, ergodic Markov system.

It is easily seen what our control decision should be if only one transition is to be made. Clearly, we should use u_1, whether the system is in state 1 or 2. We express this by

$$\mathbf{d}(1) = \begin{bmatrix} u_1 \\ u_1 \end{bmatrix} \tag{8.91}$$

With this control decision, the total cost will be

$$\min \mathbf{f}(1) = \begin{bmatrix} 6 \\ 18 \end{bmatrix} \tag{8.92}$$

Now consider the case where two successive transitions are to be made. Our decision for the first transition is to be such that the total cost for both transitions is minimized. There are four cases, as shown in Fig. 8-8. In the figure, the nodes at the left represent the current state; those at the right the possible states resulting from one transition. The numbers alongside the branches represent the applicable transition probabilities. Next to each state node is a number which indicates the cost associated with the transition out of this state. For the states at the right, for which one transition remains, we have selected the previously determined *minimum costs* (8.92), presuming that the optimum decision (8.91) would be used when these states are reached.

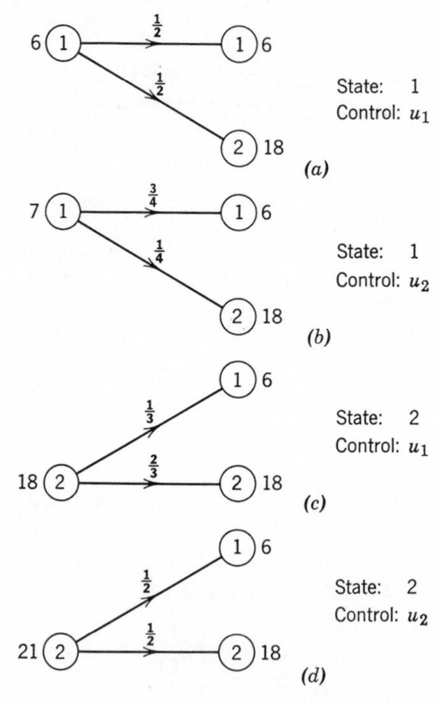

Figure 8-8 Alternatives for cost minimization over two transitions.

Let us now calculate the minimum costs for each of the four cases in Fig. 8-8. We simply take the cost of leaving the initial state plus the costs of leaving the subsequent states, the latter weighted by the appropriate transition probabilities. We find:

(*a*) State 1, Control u_1: $6 + \frac{1}{2}(6) + \frac{1}{2}(18) = 18$
(*b*) State 1, Control u_2: $7 + \frac{3}{4}(6) + \frac{1}{4}(18) = 16$
(*c*) State 2, Control u_1: $18 + \frac{1}{3}(6) + \frac{2}{3}(18) = 32$
(*d*) State 2, Control u_2: $21 + \frac{1}{2}(6) + \frac{1}{2}(18) = 33$

If we are in state 1, with two transitions to be made, u_1 will give a total cost of 18 and u_2 a cost of 16. If we are in state 2, u_1 gives a cost of 32 and

u_2 gives one of 33. Clearly, our optimum control decision for two transitions must be

$$\mathbf{d}(2) = \begin{bmatrix} u_2 \\ u_1 \end{bmatrix} \tag{8.93}$$

The resulting minimum cost will be

$$\min \mathbf{f}(2) = \begin{bmatrix} 16 \\ 32 \end{bmatrix} \tag{8.94}$$

We can now repeat the procedure to determine the minimum costs for three transitions. We can again use Fig. 8-8 to illustrate the four possible cases; we need merely replace the state costs 6 and 18 at the right, applicable when only one transition is to follow, by the state costs 16 and 32, respectively, obtained from (8.94) for the case when two transitions are to follow. The corresponding calculations then are

(*a*) State 1, Control u_1: $6 + \frac{1}{2}(16) + \frac{1}{2}(32) = 30$
(*b*) State 1, Control u_2: $7 + \frac{3}{4}(16) + \frac{1}{4}(32) = 27$
(*c*) State 2, Control u_1: $18 + \frac{1}{3}(16) + \frac{2}{3}(32) = 44.667$
(*d*) State 2, Control u_2: $21 + \frac{1}{2}(16) + \frac{1}{2}(32) = 45$

Thus

$$\mathbf{d}(3) = \begin{bmatrix} u_2 \\ u_1 \end{bmatrix} \tag{8.95}$$

and

$$\min \mathbf{f}(3) = \begin{bmatrix} 27 \\ 44.667 \end{bmatrix} \tag{8.96}$$

If we continue this procedure, we obtain the data in the following table:

k	$d_1(k)$	$d_2(k)$	$\min f_1(k)$	$\min f_2(k)$	$\min f_2(k) - \min f_1(k)$
1	u_1	u_1	6	18	12
2	u_2	u_1	16	32	16
3	u_2	u_1	27	44.67	17.67
4	u_2	u_1	38.42	56.78	18.36
5	u_2	u_2	50.02	68.60	18.58
6	u_2	u_2	61.68	80.31	18.63

We observe that as k becomes larger, the loss for each additional transition approaches a constant value of approximately 11.7, regardless of whether state 1 or state 2 was our initial state. However, if we start in state 1, the total cost will always be less by about 18.6 than if we start in state 2.

We thus note that there is a simple iterative procedure for finding the optimum control policy for a given, finite-state, ergodic system. The procedure merely requires that for each state i we find the particular value of l (corresponding to the control input u_l) that minimizes the cost function $\mathbf{f}$ in accordance with the recurrence relation:

$$\boxed{[f_i(k+1)]_{\min} = \min_l \left\{ c_i^l + \sum_{j=1}^{\nu} q_{ji,l}[f_j(k)]_{\min} \right\}} \tag{8.97}$$

$$l = 1, 2, \ldots, \alpha$$
$$i = 1, 2, \ldots, \nu$$
$$k = 0, 1, \ldots, \infty$$

The set of these l-values then yields the optimum control decision $\mathbf{d}(k+1)$.

The recurrence relation (8.97), obtained intuitively here, can also be derived directly from Bellman's *principle of optimality*[18] for multistage decision processes. According to this principle, *an optimum policy for a multistage decision process must have the property that at each stage of the process, the remaining decisions must represent an optimum policy with respect to the state resulting from the earlier decisions.* The recurrence relation (8.97) is a specific embodiment of this principle, applicable to controllable, finite-state Markov systems.

From (8.97) we can obtain the optimum control policy for operating a system for k transitions such that the mean total cost over these transitions is minimized. Occasionally, however, we will be concerned with systems that "run forever." For such systems we will be interested in the control policy that minimizes the *mean cost per transition* rather than the mean total cost starting from a particular state. We shall now describe a procedure for the solution of these "long-term" problems; the procedure is based on the work of Howard.[19]

For large values of k, the mean total cost of an ergodic Markov system operated under a particular control policy will tend to increase linearly

[18] R. Bellman, *Dynamic Programming*, Princeton University Press, Princeton, N.J., 1957, Chapter 3.

[19] R. Howard, "Studies in Discrete Dynamic Programming," Doctoral thesis, Department of Electrical Engineering, Massachusetts Institute of Technology, Cambridge, Mass., June 1958; also *Dynamic Programming and Markov Processes*, M. I. T. Press, Cambridge, Mass., 1960.

with k; that is, we can write

$$f_i(k) = kL + v_i \quad \text{for large } k, \qquad i = 1, 2, \ldots, \nu \tag{8.98}$$

where L, the mean cost increase per transition, is independent of the initial state. The quantity v_i is a constant cost, attributable to having the system start in state i. We shall refer to the equations (8.98) as the *asymptotic cost equations.*

If we substitute (8.98) in (8.77), we obtain

$$(k + 1)L + v_i = c_i + \sum_{j=1}^{\nu} q_{ji}[kL + v_j] \tag{8.99}$$

Since from (8.71)

$$\sum_{j=1}^{\nu} q_{ji} = 1 \tag{8.71}$$

we can simplify (8.99) to yield

$$L + v_i = c_i + \sum_{j=1}^{\nu} q_{ji} v_j, \qquad i = 1, 2, \ldots, \nu \tag{8.100}$$

Let us now subtract v_ν from both sides of (8.100) and again make use of (8.71):

$$L + v_i - v_\nu = c_i + \sum_{j=1}^{\nu} q_{ji} v_j - v_\nu \sum_{j=1}^{\nu} q_{ji} \tag{8.101}$$

$$= c_i + \sum_{j=1}^{\nu-1} q_{ji}(v_j - v_\nu), \qquad i = 1, 2, \ldots, \nu \tag{8.102}$$

From (8.102) we can obtain ν simultaneous, linear equations (one for each value of i) in the ν unknowns $v_1 - v_\nu, v_2 - v_\nu, \ldots, v_{\nu-1} - v_\nu$, and L. We shall refer to the $v_i - v_\nu$ as the *relative starting costs* and denote them by w_i; they represent the increases in cost for starting in state i rather than in state ν.

Thus given some constant[20] control policy,

$$\mathbf{d} = \begin{bmatrix} u_{l_1} \\ u_{l_2} \\ \cdot \\ \cdot \\ \cdot \\ u_{l_\nu} \end{bmatrix} \tag{8.103}$$

we can find the *relative* asymptotic cost equations

$$f_i' = f_i(k) - v_\nu = kL + w_i \tag{8.104}$$

[20] For a stationary, regular Markov system, a control policy designed to minimize the mean costs per transition for large k will be independent of k.

where

$$w_i = v_i - v_\nu \tag{8.105}$$

by solving the ν simultaneous equations:

$$\boxed{L + w_i = c_i^{l_i} + \sum_{j=1}^{\nu-1} q_{ji,l_i} w_j, \qquad i = 1, 2, \ldots, \nu} \tag{8.106}$$

We note, of course, that

$$w_\nu = 0 \tag{8.107}$$

Our objective here is to find the particular policy that minimizes the value L. We can accomplish this objective in the following manner. Suppose we assume some control policy (8.103). From (8.106) we can obtain the values of L and w_j, $j = 1, 2, \ldots, v - 1$, corresponding to this policy. The values of w_j thus found can then be used to find a better policy by substituting them in the expression

$$\mathscr{F}_i^{\,l} = c_i^{\,l} + \sum_{j=1}^{\nu-1} q_{ji,l} w_j \tag{8.108}$$

and minimizing this expression over l for each i. The resulting set of l-values, if different from the previous set, yields a new control policy in accordance with which (8.106) can now be re-evaluated. The procedure is repeated until the minimization of (8.108) gives no further changes in policy. At each iteration a lower value of L is obtained. When the procedure terminates, the minimum mean cost per transition, $L_{\min}$, will have been found.[21]

We shall illustrate the method by applying it to the same example for which we previously minimized the costs over a *finite* number of transitions.

Example

We again consider the two-state, controllable, ergodic system characterized by (8.89) and (8.90). We desire the optimum control policy for operating this system for an indefinite period of time, that is, *the policy that will minimize the mean cost per transition* as $k \to \infty$.

As an initial control policy, let us assume

$$\mathbf{d}^1 = \begin{bmatrix} u_1 \\ u_1 \end{bmatrix} \tag{8.109}$$

[21] For a proof of convergence, see R. Howard, *Dynamic Programming and Markov Processes*, pp. 42–43.

Then from (8.106)

$$\begin{aligned} L + w_1 &= c_1{}^1 + q_{11,1}w_1 = 6 + \tfrac{1}{2}w_1 \\ L &= c_2{}^1 + q_{12,1}w_1 = 18 + \tfrac{1}{3}w_1 \end{aligned} \tag{8.110}$$

Solving for L and w_1, we find

$$L = 13.2, \qquad w_1 = -14.4$$

This value of w_1 is now used to improve the policy. We evaluate (8.108) for every state i, using all possible values of l in each case.

For $i = 1$:

$$\mathscr{F}_1{}^1 = 6 + \tfrac{1}{2}(-14.4) = -1.2$$

$$\mathscr{F}_1{}^2 = 7 + \tfrac{3}{4}(-14.4) = -3.8 \text{ (min)}$$

For $i = 2$:

$$\mathscr{F}_2{}^1 = 18 + \tfrac{1}{3}(-14.4) = 13.2 \text{ (min)}$$

$$\mathscr{F}_2{}^2 = 21 + \tfrac{1}{2}(-14.4) = 13.8$$

The improved policy thus is

$$\mathbf{d}^2 = \begin{bmatrix} u_2 \\ u_1 \end{bmatrix} \tag{8.111}$$

With this policy, we have from (8.106)

$$\begin{aligned} L + w_1 &= c_1{}^2 + q_{11,2}w_1 = 7 + \tfrac{3}{4}w_1 \\ L &= c_2{}^1 + q_{12,1}w_1 = 18 + \tfrac{1}{3}w_1 \end{aligned} \tag{8.112}$$

Solving for L and w_1, we obtain

$$L = 11.714, \qquad w_1 = -18.857$$

We again attempt to improve the policy by using the new value of w_1 to evaluate (8.108) for all i and l. We find

For $i = 1$:

$$\mathscr{F}_1{}^1 = 6 + \tfrac{1}{2}(-18.857) = -3.428$$

$$\mathscr{F}_1{}^2 = 7 + \tfrac{3}{4}(-18.857) = -7.143 \text{ (min)}$$

For $i = 2$:

$$\mathscr{F}_2{}^1 = 18 + \tfrac{1}{3}(-18.857) = 11.714$$

$$\mathscr{F}_2{}^2 = 21 + \tfrac{1}{2}(-18.857) = 11.572 \text{ (min)}$$

The new policy thus becomes

$$\mathbf{d}^3 = \begin{bmatrix} u_2 \\ u_2 \end{bmatrix} \tag{8.113}$$

Upon evaluating (8.106) with this new policy, we find

$$L = 11.667, \qquad w_1 = -18.667$$

If this last value of w_1 is now used in the evaluation of (8.108), the resulting policy is found to be the same as (8.113). Hence the optimum policy has been found, and the minimum mean cost per transition is the last value of L, that is, 11.667.

Although we have discussed the foregoing optimization procedure only with respect to systems possessing a single ergodic state set, the method can also be applied to systems with multiple ergodic state sets if some minor modifications are made.[22]

REFERENCES

Amara, R. C., "The Linear Least Squares Synthesis of Multivariable Control Systems," *Trans. AIEE*, vol. 78, part 2, pp. 115–120, May 1959.

Barker, R. H., *The Theory of Pulse Monitored Servomechanisms and Their Use for Prediction*, Report 1046, Signals Research and Development Establishment, Ministry of Supply, Christchurch, Hants, England, November 1950.

Bellman, R., *Dynamic Programming*, Princeton University Press, Princeton, N.J., 1957.

Bellman, R., *Introduction to Matrix Analysis*, McGraw-Hill Book Co., New York, 1960.

Bellman, R., "A Markovian Decision Process," *J. Math. and Mech.*, vol. 6, no. 5, pp. 679–684, 1957.

Bellman, R. and S. E. Dreyfus, *Applied Dynamic Programming*, Princeton University Press, Princeton, N.J., 1962.

Blackwell, D., "Discrete Dynamic Programming," *Ann. Math. Stat.*, vol. 33, no. 2, pp. 719–726, June 1962.

Bode, H. W. and C. E. Shannon, "A Simplified Derivation of Linear Least Square Smoothing and Prediction Theory," *Proc. IRE*, vol. 38, pp. 417–424, 1950.

Bruce, G. D. and K. S. Fu, "A Model for Finite-State Probabilistic Systems," *Proc. First Allerton Conference Circuit System Theory*, pp. 632–651, November 15–17, 1963.

Carlyle, J. W., *Equivalent Stochastic Sequential Machines*, Series 60, Issue 415, Electronics Research Laboratory, University of California, Berkeley, November 1961.

Chang, S. S. L., "Statistical Design Theory for Strictly Digital Sampled-Data Systems," *Trans. AIEE*, vol. 76, part 1, pp. 702–709, 1957.

Chang, S. S. L., *Synthesis of Optimum Control Systems*, McGraw-Hill Book Co., New York, 1961.

Davenport, W. B. and W. L. Root, *An Introduction to the Theory of Random Signals and Noise*, McGraw-Hill Book Co., New York, 1958.

Davis, M. C., "On Factoring the Spectral Matrix," *Proc. 1963 Joint Autom. Control Conf.*, American Institute of Chemical Engineers, New York, pp. 459–467, 1963.

DeRusso, P. M., "Optimum Linear Filtering of Signals Prior to Sampling," *Trans. AIEE*, vol. 79, part 2, pp. 549–555, 1960.

[22] R. Howard, *op. cit.* pp. 60–65.

Doob, J. L., *Stochastic Processes*, John Wiley and Sons, New York, 1953.

Feller, W., *An Introduction to Probability Theory and Its Applications*, (2nd Ed.) John Wiley and Sons, New York, 1957.

Franklin, G., *The Optimum Synthesis of Sampled-Data Systems*, Doctoral dissertation, Department of Electrical Engineering, Columbia University, New York, May 1955.

Friedland, B., "Least Squares Filtering and Prediction of Nonstationary Sampled Data," *Inform. Control*, vol. 1, pp. 297–313, 1958.

Gantmacher, F. R., *Applications of the Theory of Matrices*, Interscience Publishers, New York, 1959.

Howard, R. A., *Dynamic Programming and Markov Processes*, M. I. T. Press, Cambridge, Mass., 1960.

Kalman, R., "When Is a Linear Control System Optimal?" *Proc. 1963 Joint Autom. Control Conf.*, American Institute of Chemical Engineers, New York, pp. 1–15, 1963.

Kemeny, J. G. and J. L. Snell, *Finite Markov Chains*, P. Van Nostrand Company, Princeton, N.J., 1960.

Laning, J. H. and R. H. Battin, *Random Processes in Automatic Control*, McGraw-Hill Book Co., New York, 1956.

Lee, Y. W., *Statistical Theory of Communication*, John Wiley and Sons, New York, 1960.

Newton, G. C., L. A. Gould and J. F. Kaiser, *Analytical Design of Linear Feedback Controls*, John Wiley and Sons, New York, 1957.

Parzen, E., "The Function Space Point of View in Time Series Analysis," *Proc. 1963 Joint Autom. Control Conf.*, American Institute of Chemical Engineers, New York, pp. 437–445, June 1963.

Pontryagin, L. S., V. G. Boltyanskii, R. V. Gamkrelidze, and E. F. Mishchenko, *The Mathematical Theory of Optimal Processes*, Interscience Publishers, Division of John Wiley and Sons, New York, 1962.

Silver, E. A., *Markovian Decision Processes with Uncertain Transition Probabilities or Rewards*, Doctoral dissertation, Department of Civil Engineering, Massachusetts Institute of Technology, Cambridge, Mass., 1963.

Sittler, R. W., "System Analysis of Discrete Markov Processes," *IRE Trans. Circuit Theory*, vol. CT-3, no. 1, pp. 257–266, December 1956.

Spilker, J. J., "Theoretical Bounds on the Performance of Sampled-Data Communication Systems," *IRE Trans. Circuit Theory*, vol. CT-7, no. 3, pp. 335–341, September 1960.

Stewart, R. M., "Statistical Design and Evaluation of Filters for the Restoration of Sampled Data," *Proc. IRE*, vol. 44, no. 2, pp. 253–257, February 1956.

Takacs, L., *Stochastic Processes*, Methuen and Co., London, 1960.

Tou, J. T., "Statistical Design of Linear Discrete-Data Control Systems via the Modified z-Transform Method," *J. Franklin Inst.*, vol. 271, no. 4, pp. 249–262, April 1961.

Trench, W. F., "A General Class of Discrete Time-Invariant Filters," *J. Soc. Ind. Appl. Math.*, vol. 9, no. 3, pp. 405–421, September 1961.

Wald, A., *Statistical Decision Functions*, John Wiley and Sons, New York, 1950.

Wiener, N., *The Extrapolation, Interpolation, and Smoothing of Stationary Time Series*, M.I.T. Press, Cambridge, Mass., 1949.

Zadeh, L. A., and J. R. Ragazzini, "An Extension of Wiener's Theory of Prediction," *J. Appl. Phys.*, vol. 21, pp. 645–655, July 1950.

APPENDIX A

Numerical Method for $\mathcal{Z}$-Transform Inversion

The long-division inversion of rational z transforms to obtain the values $f(k)$ of the corresponding k-domain sequences can be facilitated by means of the following procedure. The approach is particularly useful if the computation is to be performed on a desk calculator or programmed for a digital computer.

Consider the canonical form for a z transform expression

$$F(z) = \frac{a_0 z^0 + a_1 z^{-1} + \cdots + a_p z^{-p}}{b_0 z^0 + b_1 z^{-1} + \cdots + b_q z^{-q}}, \qquad |z| > R_c \tag{A.1}$$

where $b_0 \neq 0$. Division of denominator into numerator gives

$$F(z) = \sum_{k=0}^{\infty} f(k) z^{-k} \tag{A.2}$$

where the $f(k)$ are the desired values of the corresponding time sequence.

If we perform the actual division indicated by (A.1), the following results are obtained.

$$\begin{aligned} f(0) &= \frac{a_0}{b_0} \\ f(1) &= \frac{a_1 - b_1 f(0)}{b_0} \\ f(2) &= \frac{a_2 - b_2 f(0) - b_1 f(1)}{b_0} \qquad \text{etc.} \end{aligned} \tag{A.3}$$

The general term is given by

$$f(k) = \frac{a_k}{b_0} - \sum_{j=1}^{q} \frac{b_j}{b_0} f(k-j) \tag{A.4}$$

where $a_k = 0$ for $k > p$. (If $k < q$, the upper limit should be k rather than q.)

Equation (A.4) is a convenient recursion formula for the kth term of the sequence. Substituting

$$u_k = \frac{a_k}{b_0} \qquad v_j = -\frac{b_j}{b_0} \tag{A.5}$$

we obtain

$$\boxed{f(k) = u_k + \sum_{j=1}^{q} v_j f(k-j)} \tag{A.6}$$

This can be written in matrix form:

$$[f(0) \;\; f(1) \;\; f(2) \ldots f(k) \ldots \to \infty]$$
$$= [u_0 \;\; u_1 \ldots u_p \;\; 0 \ldots \to \infty] + [v_1 \;\; v_2 \ldots v_q]$$
$$\times \begin{bmatrix} 0 & f(0) & f(1) & f(2) \ldots f(q-1) & f(q) & \ldots f(k-1) \ldots \\ 0 & 0 & f(0) & f(1) \ldots f(q-2) & f(q-1) \ldots & f(k-2) \ldots \\ \cdot & \cdot & \cdot & f(0) \ldots f(q-3) & f(q-2) \ldots & f(k-3) \;\; \cdot \\ \cdot & & & \cdot & \cdot & \cdot \quad \cdot \\ \cdot & & & \ldots \;\; \cdot & \cdot & \ldots \;\; \cdot \quad \cdot \\ \cdot & & & & \cdot & \cdot \\ 0 & 0 & 0 & 0 \ldots \;\; f(0) & f(1) & \ldots f(k-q) \ldots \end{bmatrix} \to \infty \tag{A.7}$$

The right-hand side expression of (A.7) consists of the sum of a row vector **u** and the product of a row vector **v** times the semi-infinite matrix **F**. Symbolically we may write

$$\mathbf{f} = \mathbf{u} + \mathbf{vF} \tag{A.8}$$

The matrix form of (A.7) suggests the worksheet arrangement shown in Fig. A-1. At the start of the computations, the u's and v's are inserted in the form as indicated. Each new $f(k)$ is computed by summing all the values in the kth column, rows 1 through $q + 1$ inclusive. The value of $f(k)$ is then entered in the space in row 0 of column k. The products $v_j f(k)$ are then obtained and inserted in row j and column $k + j$, for $j = 1, 2, \ldots q$, as shown. The process is repeated until the desired number of $f(k)$ are obtained. The process thus consists of alternately obtaining the algebraic sum of the terms in a column and inserting simple product terms in a diagonal set of boxes on the worksheet.

		0	1	2	3	4	......	k
0		$f(0)$	$f(1)$	$f(2)$	$f(3)$	$f(4)$	......	$f(k)$
1	v_1	0	$v_1f(0)$	$v_1f(1)$	$v_1f(2)$	$v_1f(3)$	......	$v_1f(k-1)$
2	v_2	0	0	$v_2f(0)$	$v_2f(1)$	$v_2f(2)$	......	$v_2f(k-2)$
3	v_3	0	0	0	$v_3f(0)$	$v_3f(1)$	......	$v_3f(k-3)$
4	v_4	0	0	0	0	$v_4f(0)$	......	$v_4f(k-4)$
⋮	⋮	⋮	⋮	⋮	⋮	⋮	⋮ ⋮	⋮
q	v_q	0	0	0	0	0	......	$v_qf(k-q)$
$q+1$	u	u_0	u_1	u_2	u_3	u_4	......	u_k

Figure A-1 Worksheet form for z-transform inversion.

Example

Given

$$F(z) = \frac{0.16216 - 0.16216z^{-1} + 0.05405z^{-2}}{1 - 2.40540z^{-1} + 1.94595z^{-2} - 0.54054z^{-3}}$$

Hence

$$u_0 = +0.16216 \qquad v_1 = +2.4054$$
$$u_1 = -0.16216 \qquad v_2 = -1.94595$$
$$u_2 = +0.05405 \qquad v_3 = +0.54054$$

Performing the calculations as indicated in Fig. A-1, we have

$$f(0) = 0.16216 \qquad f(3) = 0.33378$$
$$f(1) = 0.22790 \qquad f(4) = 0.36817$$
$$f(2) = 0.28669 \qquad f(5) = 0.39107 \quad \text{etc.}$$

		0	1	2	3	4	5	6
		+0.16216	+0.22790	+0.28669	+0.33378	+0.36817	+0.39107	+0.40467
1	+2.4054	0	+0.39006	+0.54819	+0.68960	+0.80286	+0.88560	+0.94068
2	−1.94595	0	0	−0.31555	−0.44347	−0.55788	−0.64950	−0.71644
3	+0.54054	0	0	0	+0.08765	+0.12319	+0.15497	+0.18043
4	u	+0.16216	−0.16216	+0.05405	0	0	0	0

Figure A-2 Illustration of method for z-transform inversion.

APPENDIX B

$\mathcal{Z}$-Transform Tables

B.1 TABLE OF TRANSFORM-OPERATION EQUIVALENCES

This table gives a summary of corresponding functional operations in the discrete-time domain and the z-transform domain. The time sequences $\{f(k)\}$ are assumed to vanish for $k < 0$, and, unless otherwise indicated, the z transforms converge for all $|z| > R_f$, where R_f is the radius of convergence of $F(z)$. (*Note:* k represents discrete time, and m and l are positive integers.)

	k-Domain	z-Domain
B.1	$f(k)$	$F(z) = \sum_{k=0}^{\infty} f(k)z^{-k}$ $|z| > R_f$
B.2	$f(k + m)$	$z^m[F(z) - f(0) - z^{-1}f(1) - \cdots - z^{-m+1}f(m - 1)]$
B.3	$f(k - m)$	$z^{-m}F(z)$
B.4	$\Delta f(k)$	$(z - 1)F(z) - zf(0)$
B.5	$\Delta^2 f(k)$	$(z - 1)^2F(z) - z(z - 2)f(0) - zf(1)$
B.6	$\Delta^3 f(k)$	$(z - 1)^3F(z) - z(z^2 - 3z + 3)f(0) - z(z - 3)f(1) - zf(2)$
B.7	$\nabla f(k)$	$(1 - z^{-1})F(z)$
B.8	$\nabla^l f(k)$	$(1 - z^{-1})^l F(z)$
B.9	$a^k f(k)$	$F(a^{-1}z)$ $|z| > |a|\, R_f$
B.10	$k^l f(k)$	$\left(-z \frac{d}{dz}\right)^l F(z)$

B.1 TABLE OF TRANSFORM-OPERATION EQUIVALENCES—(*contd.*)

	k-Domain	z-Domain
B.11	$af(k) + bh(k)$	$aF(z) + bH(z)$ $\lvert z\rvert > \max(R_f, R_h)$
B.12	$f(k)h(k)$	$\frac{1}{2\pi j}\oint \zeta^{-1}H(\zeta)F(\zeta^{-1}z)\,d\zeta$ $\lvert z\rvert > R_f R_h$
B.13	$\sum_{j=0}^{\infty} f(j)h(k-j)$ $\sum_{j=0}^{\infty} h(j)f(k-j)$	$F(z)H(z)$ $\lvert z\rvert > \max(R_f, R_h)$

B.2 TABLE OF $\mathcal{Z}$-TRANSFORM FORMULAS

B.14

$$f(k) = \frac{1}{2\pi j}\oint z^{k-1}F(z)\,dz$$
$$= \sum \text{residues of } z^{k-1}F(z) \text{ at poles of } z^{k-1}F(z) \text{ for } k \geq 0$$

B.15

$$F(z) = \frac{1}{2\pi j}\oint \frac{F_s(\lambda)}{1 - z^{-1}e^{\lambda T}}\,d\lambda$$
$$= \sum \text{residues of } \frac{F_s(\lambda)}{1 - z^{-1}e^{\lambda T}} \text{ at poles of } F_s(\lambda)$$

(*Note:* $F_s(\lambda) = \mathscr{L}[f(t)]$)

B.16

$$F(z)\big|_{z=e^{sT}} = \frac{1}{T}\sum_{l=-\infty}^{\infty} F_s\left(s + \frac{2\pi jl}{T}\right)$$

B.17

Residue of $z^{k-1}F(z)$ at lth-order pole $z = a$

$$= \lim_{z\to a}\frac{1}{(l-1)!}\frac{d^{l-1}}{dz^{l-1}}(z-a)^l z^{k-1}F(z)$$

B.18

Residue of $z^{k-1}\dfrac{A(z)}{B(z)}$ at simple pole $z = a$

$$= \lim_{z\to a}\frac{A(z)}{B'(z)}$$

B.2 TABLE OF $\mathcal{Z}$-TRANSFORM FORMULAS—(*contd.*)

B.19	$f(0) = \lim_{z \to \infty} F(z)$	$(R_f < \infty)$
B.20	$\lim_{k \to \infty} f(k) = \lim_{z \to 1} (z - 1)F(z)$	$(R_f \leq 1)$
B.21	$\sum_{k=0}^{\infty} f(k) = F(z)\big\|_{z=1}$	$(R_f < 1)$
B.22	$\sum_{k=0}^{\infty} kf(k) = -\frac{d}{dz} F(z)\big\|_{z=1}$	$(R_f < 1)$
B.23	$\sum_{k=0}^{\infty} [f(k)]^2 = \mathcal{Z}^{-1}[F(z)F(z^{-1})]_{k=0}$	$(R_f < 1)$

B.3 TABLE OF $\mathcal{Z}$-TRANSFORM PAIRS

The left-hand column of this table lists the time sequences $\{f(kT)\}$ obtained by sampling continuous functions $\{f(t)\}$ at $t = kT, k = 0, 1, \ldots, \infty$. The center column gives the corresponding z transforms. For time sequences with unity spacing, we need merely let $T = 1$ in the expressions for $F(z)$. The column at the right lists the Laplace transforms of the continuous-time functions $\{f(t)\}$. If the entry in the left-hand column is normally associated only with time sequences $\{f(k)\}$, rather than samples of continuous functions $\{f(t)\}$, the right-hand entry is omitted.

Unless otherwise stated, k may take on all integer values $k \geq 0$. The z transforms converge for all $|z| > R_f$, where R_f is the radius of convergence of $F(z)$.

	$\{f(kT)\}$	$F(z)$	$\mathcal{L}[f(t)]$
B.24	$\delta(t)$	1	1
B.25	1	$\frac{z}{z-1}$	$\frac{1}{s}$
B.26	kT	$\frac{Tz}{(z-1)^2}$	$\frac{1}{s^2}$
B.27	$(kT)^2$	$\frac{T^2z(z+1)}{(z-1)^3}$	$\frac{2!}{s^3}$
B.28	$(kT)^3$	$\frac{T^3z(z^2+4z+1)}{(z-1)^4}$	$\frac{3!}{s^4}$

B.3 TABLE OF $\mathcal{Z}$-TRANSFORM PAIRS—(*contd.*)

	$\{f(kT)\}$	$F(z)$	$\mathscr{L}[f(t)]$
B.29	$(kT)^4$	$\dfrac{T^4z(z^3+11z^2+11z+1)}{(z-1)^5}$	$\dfrac{4!}{s^5}$
B.30	$(kT)^l$	$T^l\left(-z\dfrac{d}{dz}\right)^l\dfrac{z}{z-1}$	$\dfrac{l!}{s^{l+1}}$
B.31	a^k	$\dfrac{z}{z-a}$	
B.32	e^{-bTk}	$\dfrac{z}{z-e^{-bT}}$	$\dfrac{1}{s+b}$
B.33	kTe^{-bTk}	$\dfrac{Te^{-bT}z}{(z-e^{-bT})^2}$	$\dfrac{1}{(s+b)^2}$
B.34	$(kT)^2e^{-bTk}$	$\dfrac{T^2e^{-bT}z(z+e^{-bT})}{(z-e^{-bT})^3}$	$\dfrac{2}{(s+b)^3}$
B.35	$(kT)^le^{-bTk}$	$T^l\left(-z\dfrac{d}{dz}\right)^l\dfrac{z}{z-e^{-bT}}$	$\dfrac{l!}{(s+b)^{l+1}}$
B.36	$1-e^{-bTk}$	$\dfrac{(1-e^{-bT})z}{(z-1)(z-e^{-bT})}$	$\dfrac{b}{s(s+b)}$
B.37	$bkT + e^{-bTk} - 1$	$\dfrac{(bT+e^{-bT}-1)z^2+(1-e^{-bT}-bTe^{-bT})z}{(z-1)^2(z-e^{-bT})}$	$\dfrac{b^2}{s^2(s+b)}$
B.38	$k^{(l)}$*	$\dfrac{l!z}{(z-1)^{l+1}}$	
B.39	$\cos\omega kT$	$\dfrac{z(z-\cos\omega T)}{z^2-2z\cos\omega T+1}$	$\dfrac{s}{s^2+\omega^2}$
B.40	$\sin\omega kT$	$\dfrac{z\sin\omega T}{z^2-2z\cos\omega T+1}$	$\dfrac{\omega}{s^2+\omega^2}$
B.41	$e^{-bTk} \times \cos\omega kT$	$\dfrac{z(z-e^{-bT}\cos\omega T)}{z^2-2ze^{-bT}\cos\omega T+e^{-2bT}}$	$\dfrac{s+b}{s^2+2bs+(b^2+\omega^2)}$
B.42	$e^{-bTk} \times \sin\omega kT$	$\dfrac{ze^{-bT}\sin\omega T}{z^2-2ze^{-bT}\cos\omega T+e^{-2bT}}$	$\dfrac{\omega}{s^2+2bs+(b^2+\omega^2)}$
B.43	$\cosh akT$	$\dfrac{z(z-\cosh aT)}{z^2-2z\cosh aT+1}$	$\dfrac{s}{s^2-a^2}$
B.44	$\sinh akT$	$\dfrac{z\sinh aT}{z^2-2z\cosh aT+1}$	$\dfrac{a}{s^2-a^2}$

* $k^{(l)} = k(k-1)(k-2)\ldots(k-l+1)$.

B.3 TABLE OF $\mathcal{Z}$-TRANSFORM PAIRS—(*contd.*)

	$\{f(kT)\}$	$F(z)$	$\mathcal{L}[f(t)]$
B.45	$\frac{1}{k}$ $(k \geq 1)$	$-\log(1 - z^{-1})$	
B.46	$\frac{1}{k!}$	$e^{1/z}$	
B.47	$1_{(t > -\delta T)}$	$\frac{z}{z-1}$	$\frac{e^{\delta Ts}}{s}$ $0 \leq \delta < 1$
B.48	$kT + \delta T$	$\frac{\delta T z^2 + z(T - \delta T)}{(z-1)^2}$	$\frac{e^{\delta Ts}}{s^2}$
B.49	$e^{-b(kT+\delta T)}$	$\frac{ze^{-b\delta T}}{z - e^{-bT}}$	$\frac{e^{\delta Ts}}{s+b}$
B.50	$(kT + \delta T) \times e^{-b(kT+\delta T)}$	$\frac{Tze^{-b\delta T}[\delta z + (1-\delta)e^{-bT}]}{(z - e^{-bT})^2}$	$\frac{e^{\delta Ts}}{(s+b)^2}$
B.51	$1 - e^{-b(kT+\delta T)}$	$\frac{z^2(1 - e^{-b\delta T}) + z(e^{-b\delta T} - e^{-bT})}{(z-1)(z - e^{-bT})}$	$\frac{be^{\delta Ts}}{s(s+b)}$
B.52	$\cos\omega(kT + \delta T)$	$\frac{z^2\cos\omega\delta T - z\cos(\omega T - \omega\delta T)}{z^2 - 2z\cos\omega T + 1}$	$\frac{se^{\delta Ts}}{s^2 + \omega^2}$
B.53	$\sin\omega(kT + \delta T)$	$\frac{z^2\sin\omega\delta T + \sin(\omega T - \omega\delta T)}{z^2 - 2z\cos\omega T + 1}$	$\frac{\omega e^{\delta Ts}}{s^2 + \omega^2}$
B.54	$e^{-b(kT+\delta T)} \times \cos\omega(kT + \delta T)$	$\frac{ze^{-bT\delta}[z\cos\omega\delta T - e^{-bT}\cos(\omega T - \omega\delta T)]}{z^2 - 2ze^{-bT}\cos\omega T + e^{-2bT}}$	$\frac{(s+b)e^{\delta Ts}}{s^2 + 2bs + (b^2 + \omega^2)}$
B.55	$e^{-(bk+\delta T)} \times \sin\omega(kT + \delta T)$	$\frac{ze^{-bT\delta}[z\sin\omega\delta T + e^{-bT}\sin(\omega T - \omega\delta T)]}{z^2 - 2ze^{-bT}\cos\omega T + e^{-2bT}}$	$\frac{\omega e^{\delta Ts}}{s^2 + 2bs + (b^2 + \omega^2)}$

Note: $0 \leq \delta < 1$.

Problems

CHAPTER 1

1.1 Indicate which, if any, of the coefficients in the following equation *must* be zero for the equation to describe a system that is

(*a*) linear
(*b*) stationary
(*c*) nonanticipatory

Assume that u and y represent the input and output, respectively, and that all the c's are constants.

$$c_9 \frac{d^3y}{dt^3} + c_8\left(\frac{dy}{dt}\right)^2 + (c_7 + c_6 t + c_5 y + c_4 y^{c_3 t})\frac{dy}{dt} + c_2 y = c_1 u$$

1.2 Let E be the advance operator, defined by $Ef(k) = f(k+1)$, and let $P(E)$ be a polynomial in this operator.

$$P(E) = a_n E^n + a_{n-1}E^{n-1} + \cdots + a_1 E + a_0$$

Determine $P(E)b^k$, where b is a constant.

1.3 The forward-difference operator Δ is defined by $\Delta f(k) = f(k+1) - f(k)$. Express the forward difference of a product, $\Delta f(k)g(k)$ in terms of the forward differences $\Delta f(k)$ and $\Delta g(k)$.

1.4 The summation operator Σ is defined as the right inverse of the forward difference operator Δ. Thus for $\Delta f(k) = f(k+1) - f(k)$,

$$\Delta \sum f(k) = f(k)$$

but

$$\sum \Delta f(k) = f(k) + c$$

where c is a constant.

(*a*) Prove that $\Sigma\, a^k = \dfrac{a^k}{a-1} + c$

(*b*) Determine $\Sigma \cos \omega k$

CHAPTER 2

2.1 A linear system possesses a weighting sequence

$$h(k) = (k-1)e^{-a(k-1)}$$

where $a > 0$ and $k = 1, 2, \ldots, \infty$.

(*a*) Is the system nonanticipatory?
(*b*) Is it stationary?

2.2 The weighting sequence of a linear system is given by $2e^{-3(k-1)T}$, for $k \geq 1$. Let the input to the system be $u(k) = 4kT$ for all integers $k \geq 0$. $T = 0.1$ sec.

(*a*) Find a series expression for the system output sequence.

(*b*) What is the output at $t = 0.3$ sec?

(*c*) Find a *closed-form* expression for the output sequence.

2.3 A system has the weighting sequence $\{1, 3, 2, 1\}$ associated, respectively, with $k = 1, 2, 3, 4$. Determine a closed-form expression for the output sequence for all $k \geq 1$, given that the input sequence is 2^k for $k \geq 0$.

2.4 The sequence $(\frac{1}{2})^{|k|}$ for *all* integers k is impressed upon a system having a weighting sequence that consists only of the following three nonzero terms: $h(1) = 4$, $h(2) = -12$, and $h(3) = 5$. Determine the output sequence for all integers k.

2.5 Experiments are performed to develop a description for a given system of unknown characteristics. It is observed that if an input $u(k) = 3(2)^{k-i}$ is applied to the system starting at time i with the system fully relaxed, the resulting output for $k \geq i$ is given by:

$$y(k) = 3k^{-1}(2)^{k-i} - k^{-1}(3)^{k-i+1} + (k - i)k^{-1}(3)^{k-i}$$

Under the assumption that the system is linear, determine the weighting sequence for the system.

2.6 A system is described by the following transmission matrix:

$$\begin{bmatrix} y(1) \\ y(2) \\ y(3) \\ \cdot \\ \cdot \\ y(k) \end{bmatrix} = \begin{bmatrix} 480 & 0 & 0 & \dots & 0 \\ 240 & 120 & 0 & \dots & 0 \\ 120 & 60 & 30 & \dots & 0 \\ \cdot & \cdot & \cdot & \dots & 0 \\ & & & & \\ \cdot & \cdot & \cdot & \dots & \cdot \end{bmatrix} \begin{bmatrix} u(0) \\ u(1) \\ u(2) \\ \cdot \\ \cdot \\ u(k-1) \end{bmatrix}$$

(*a*) What is the output at time $k = 4$ if $u(0) = 1$, $u(1) = 2$, $u(2) = 4$, $u(3) = 2$, $u(4) = 5$, and $u(5) = 1$?

(*b*) Is the system linear?

(*c*) Is the system stationary?

(*d*) Describe the system in terms of a convolution summation.

2.7 The system

$$y(k) - 6y(k-1) + 11y(k-2) - 6y(k-3) = u(k-1) - 4u(k-2)$$

has impressed upon it the input $u(k) = 4^k$ for $k \geq 0$. It is known that $y(0) = 2$, $\Delta y(0) = 1$, and $\Delta^2 y(0) = 2$.

(*a*) Represent the system in terms of state equations. (Do not use normal coordinates.)

(*b*) Determine the state vector $\mathbf{x}(0)$ for the system representation selected in (*a*).

(*c*) Determine $y(3)$ and $y(4)$.

(*d*) Derive the transformation required to transform from the coordinates used in (*a*) to *normal coordinates*, and represent the system in normal form.

(*e*) Find the initial-state vector for (*d*).

(*f*) Draw a signal flow graph for the normal-form representation (*d*), showing the input terminal at the left and the output terminal at the right.

(*g*) Represent the system in terms of a convolution summation.

2.8 A system is characterized by the following matrices:

$$\mathbf{A} = \begin{bmatrix} 10 & 1 & 0 \\ -34 & 0 & 1 \\ 40 & 0 & 0 \end{bmatrix} \qquad \mathbf{B} = \begin{bmatrix} 2 & -5 \\ -12 & 30 \\ 20 & -50 \end{bmatrix}$$

$$\mathbf{C} = \begin{bmatrix} 4 & 1 & \frac{1}{4} \\ 8 & 2 & \frac{2}{5} \end{bmatrix}$$

(*a*) Determine whether the system is completely controllable.

(*b*) Determine whether the system is completely observable.

(*c*) Determine the weighting sequence matrix of the system.

(*d*) Draw a signal flow graph for the system, using the normal-coordinate form.

2.9 Determine the output $y(k)$ for all $k \geq 1$ for the system described by the difference equation:

$$y(k+3) - 7y(k+2) + 16y(k+1) - 12y(k) = u(k+2)$$

where $u(k) = 0$ for all $k \geq 0$, $y(1) = 0$, $y(2) = 1$, and $\Delta y(2) = 7$.

2.10 Consider the system shown in the accompanying figure.

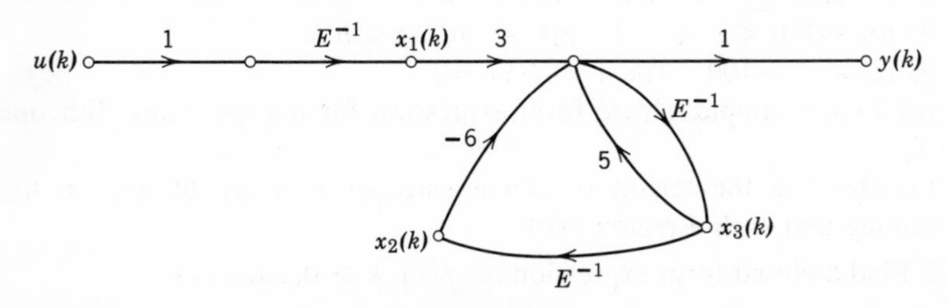

Problem 2.10

(*a*) Is the system completely controllable and completely observable?
(*b*) Find the input-to-output weighting sequence for this system.

CHAPTER 3

3.1 (*a*) Determine $\mathcal{Z}[k^3]$, $k \geq 0$
(*b*) Determine $\mathcal{Z}[k^{(3)}]$, $k \geq 0$
Note: $k^{(i)} = k(k-1)(k-2)\ldots(k-i+1)$
(*c*) Determine $\mathcal{Z}[e^{ak}\cosh \omega k]$, $k \geq 0$.

3.2 Determine $\mathcal{Z}[T^2k^2a^{bkT}\sin \omega k]$, $k \geq 0$, where T, a, b, and ω are constants. Indicate the region of convergence.

3.3 Find closed-form expressions for the z transforms of each of the following sequences. Let the first term correspond to $k = 1$ in each case. Indicate the regions of convergence.
(*a*) 0, 1×2, 2×3, 3×4, ... ∞.
(*b*) 3, $2(3)^2$, $3(3)^3$, $4(3)^4$, ... ∞.
(*c*) 1, 1, 2, 3, 5, 8, 13, 21, 34, ... ∞.

3.4 Find the generating function of the function

$$f(k) = \left(\frac{1}{5}\right)^{|k|} + \left(\frac{1}{3}\right)^{|k|}$$

for $-\infty \leq k \leq \infty$. Express the result as a rational function in z and indicate the region of convergence.

3.5 Determine the generating function of f in closed form, given that

$$\begin{aligned} f(k) &= -k3^k && \text{for } k < 0 \\ &= 5 && \text{for } k = 0 \\ &= 2^{-k} && \text{for } k > 0 \end{aligned}$$

3.6 A time sequence f is described as follows:
1. at $k = -3$, $f(k) = A$
2. for all $k \geq -3$, $f(k+1)/f(k) = a$
3. for $-100 \leq k < -3$, $f(k+1)/f(k) = b^{-1}$
4. for $k < -100$, $f(k+1)/f(k) = c^{-1}$

(*a*) Find a simple, closed-form expression for the generating function of f.
(*b*) Describe the region of convergence of $F(z)$ as obtained in (*a*), assuming that such a region exists.

3.7 Find a closed-form expression for $f(k)$, $k \geq 0$, such that

$$\mathcal{Z}[f(k)] = \tan^{-1} z^{-1}, \quad |z| > 1$$

3.8 A discrete-time system transfer function is given as

$$H(z) = \frac{6(2 - z^{-1})}{1 - z^{-1} - 2z^{-2}} \qquad \text{for} \quad |z| > 2$$

The input consists of

$$u(0) = 1, \quad u(1) = -1, \quad u(2) = -1, \quad u(3) = 1$$

Determine the output $y(k)$ for $k \geq 0$.

3.9 A linear, stationary, discrete-time system has the transfer function

$$H(z) = 2 + \tfrac{2}{3}z^{-1} - 20z^{-2}$$

defined for all z except $|z| = 0$. The input to the system is

$$u(k) = \left(\frac{1}{3}\right)^{|k|}$$

for all integers k. Find the output $y(k)$ of the system for *all* k.

3.10 Given that

$$F(z) = \frac{z^2 - z - 27 + 45z^{-1}}{z^2 - 8z + 15}$$

converges only within the annulus bounded by $|z| = 3$ and $|z| = 5$, find $f(k)$ for all k.

3.11 Express $f(k)$ in a simple, compact form, given that

$$F(z) = \frac{75z^2 + 9.82z}{25z^2 - 5z + 1} \qquad \text{for} \quad |z| > \frac{1}{5}$$

3.12 Determine $f(k)$ for all k, given that

$$F(z) = \frac{3z^4 - 28z^3 + 62z^2 + 15z + 50}{z^3 - 10z^2 + 25z}$$

and converges for $|z| < 5$.

3.13 Determine $f(k)$, given that

$$F(z) = \frac{2z^{103} - 17z^{102} + 40z^{101} - 24z^{100} + 5z^3 - 45z^2 + 130z - 120}{z^{103} - 9z^{102} + 26z^{101} - 24z^{100}}$$

for $3 < |z| < 4$.

3.14 Given

$$F(z) = \frac{4z^3 - 35z^2 + 100z - 108}{z^3 - 10z^2 + 33z - 36}$$

for $|3| < z < |4|$.

(*a*) Determine $f(k)$ by using the method of *partial-fraction expansion.*
(*b*) Determine $f(k)$ by using the inversion integral.

CHAPTER 4

4.1 Determine the z transforms of the time sequences that would be obtained by sampling at $t = kT$ the time functions possessing the following Laplace transforms:

(*a*) $\dfrac{A}{s(s+a)}$

(*b*) $\dfrac{A}{(s+a)(s+b)^2}$

(*c*) $\dfrac{A}{s^2(s+a)^2}$

(*d*) $\dfrac{5(s+2)}{s^2+2s+3}$, $\quad T = 0.1$

4.2 The input to a digital integrator is obtained by sampling the function

$$f(t) = 1 - 2e^{-3t} + e^{-2t}, \qquad t \geq 0$$

at a rate of once every 0.05 sec. The integrator works on the basis of Simpson's $\frac{2}{3}$ rule,

$$\int_0^{kT} f(t)\,dt \approx \frac{2T}{3}\,[\tfrac{1}{2}f(0)$$

$$+ 2f(T) + f(2T) + 2f(3T) + f(4T) + \cdots + 2f(kT - T) + \tfrac{1}{2}f(kT)]$$

(where k = even) and was initially cleared.

(*a*) Using z-transform techniques, obtain a closed expression for the output $y(kT)$.

(*b*) Develop a generalized expression for the error between $y(kT)$ and the time integral of $f(t)$ at $t = kT$.

4.3 A signal $f(t)$ has a Fourier transform $F(j\omega)$ such that

$$F(j\omega) = 0 \quad \text{for} \quad |\omega| > \Omega/2$$

The signal is used to modulate a carrier $\cos 5\Omega t$. Let the modulated carrier be denoted by $e(t)$. Derive the reconstruction formula which gives $f(t)$ in terms of a convolution summation of the samples $e(kT)$ with a filter weighting function ($\Omega = 2\pi/T$).

4.4 A signal $f(t)$ has a Fourier transform which vanishes for $|\omega| > \Omega/2$. ($\Omega = 2\pi/T$.)

(*a*) Show that $f(t)$ can be uniquely determined from a knowledge of the following (for all integers k):

$$f(4kT),\quad f'(4kT),\quad f(4kT+\tau),\quad f'(4kT+\tau)$$

(Note: $0 < \tau < 4T$.)

(*b*) Find the required reconstruction formula.

4.5 Determine the z transform of the sequence obtained by sampling the time function

$$f(t) = (t - 3.7T)e^{-5(t-3.7T)}$$

at $t = 0.2k$, $k = 0, 1, 2, \ldots, \infty$.

4.6 A delta-function series $\sum_{k=0}^{\infty} \delta(t - 6k - 1)$ has been modulated by the continuous-time function $f(t) = te^{-t}$ The Laplace transform of the modulated delta-function series was obtained and then converted into a closed-form z transform expression by the substitution $z = e^s$. What is this z transform?

4.7 At $t = 0$ a signal $e^{-at} \sin \omega t$ is applied to a perfect delay line of time delay $0.1T$ sec. The output of the delay line is sampled at $t = kT$, $k = 0, 1, \ldots, \infty$. Determine the z transform of the resulting time sequence.

4.8 A function $\{f(t)\}$ is zero for $t < 0$ and has a Laplace transform given by

$$F(s) = \frac{(2+a)s + 2a}{s^2(s+a)}.$$

(*a*) Find the z transform of $\{f(kT - v)\}$, where $v = nT - mT$, n is a positive integer, and m is a positive constant less than unity.

(*b*) What is the region of convergence of this z transform?

4.9 A 10-volt, 0.1-sec-wide rectangular pulse is applied at $t = 0$ to the circuit shown in the accompanying figure. The output of the circuit is

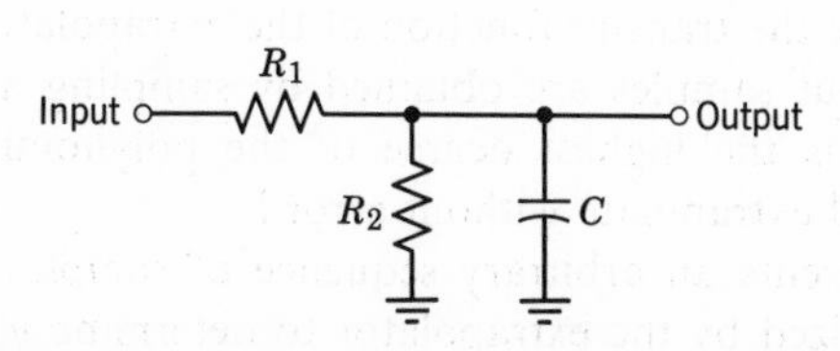

Problem 4.9 $R_1 = 2$ megohms, $R_2 = 5$ megohms, $C = 1$ microfarad.

read automatically by means of a vacuum tube voltmeter at intervals of 0.02 sec. The resulting reading $u(k)$ is transmitted to a high-speed digital computer, which performs the following operation on the data:

$$y(k) = au(k) + (1 - a)y(k - 1)$$

where $y(k)$ is the quantity calculated by the computer, $a = 0.4$, and the computer requires 0.01 sec to carry out the computation; that is, an output sample becomes available 0.01 sec after an input sample was supplied to the computer.

Using z-transform techniques, determine the expression for the computer *output* if the *first* reading of the meter takes place at $t = 0.005$ sec. Indicate the exact times at which computer outputs become available.

CHAPTER 5

5.1 Determine the transfer function of the interpolator which interpolates the function f in the interval $kT - T \leq t \leq kT$ on the basis of the values of $f(kT)$ and $f(kT - T)$.

5.2 Determine the transfer function of a Lagrangian extrapolator which extrapolates the function f in the interval $kT \leq t < kT + T$ on the basis of the values of $f(kT), f(kT - T), f(kT - 2T)$, and $f(kT - 3T)$.

5.3 A smoothing extrapolator is described by the following three equations:

$$y(t) = y(kT) + \left(\frac{t - kT}{T}\right)r(kT)$$

$$y(kT) = au(kT) + (1 - a)[y(kT - T) + r(kT)]$$

$$r(kT) = br(kT - T) + (1 - b)[y(kT - T) - y(kT - 2T)]$$

where y is the smoothed, extrapolated output, u represents the "noisy" input samples, and r is a smoothed measure of the discrete rate of change of u. The constants a and b are smoothing parameters which lie between 0 and 1.

(*a*) Determine the transfer function of the extrapolator.

(*b*) If the input samples are obtained by sampling a true polynomial function, what is the highest degree of the polynomial for which the extrapolator will extrapolate without error?

(*c*) If u represents an arbitrary sequence of samples, how many past samples are utilized by the extrapolator to determine $y(t)$ in a particular interval?

5.4 A so-called *modified first-order hold extrapolator* extrapolates from a sequence of uniformly spaced values $f(kT)$ in accordance with the following rule:

$$f(kT + v) = f(kT) + a[f(kT) - f(kT - T)]v/T$$

where $0 \leq v < T$ and $0 \leq a \leq 1$. (Note that if $a = 0$, we have a zero-order hold, and if $a = 1$, we have a true first-order hold.)

Derive the transfer function that characterizes this extrapolator.

CHAPTER 6

6.1 Find the z transform of $h(t)$ for $T = 0.2$ given that the z transform for $T = 0.4$ is

(*a*)
$$H(z) = \frac{6 + 0.21z^{-1} + 0.39z^{-2}}{(1 - 0.5z^{-1})(1 - 0.8z^{-1})^2}$$

(*b*)
$$H(z) = \frac{3 - 0.86z^{-1} + 0.5283z^{-2}}{(1 - 0.36z^{-1})(1 - 0.81z^{-1})^2}$$

6.2 A continuous-time system characterized by the transfer function $\dfrac{5}{s(s+3)}$ has applied to it the delta-function series $2e^{-4kT}$, where $T = 0.2$ sec ($k = 0, 1, 2, \ldots \rightarrow \infty$).

(*a*) Find the system output at $t = kT$.
(*b*) Find the system output at $t = kT/3$.

6.3 A sequence of flat-topped pulses of amplitude 9^k are applied to the system shown in the accompanying figure at $t = kT$, $k = 0, 1, 2, \ldots, \infty$. The pulses are all of uniform width $T/2$. Note: $e^{-aT/2} = 0.5$.

What is the output of the system at the instants $t = kT$, that is, what is $y(kT)$?

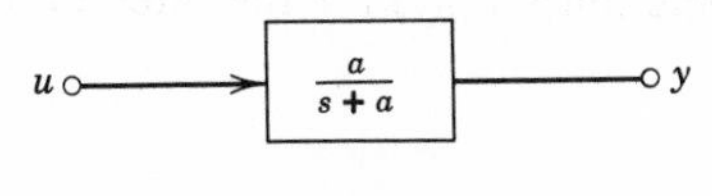

Problem 6.3

6.4 The system shown in the accompanying figure has two sampling operators operating with different periods. The samplers are synchronized at $t = 0$. The input $u(t)$ is given by $10e^{-bt}$.

(*a*) Find $Y(z)$ for the case where $T_1 = T$ and $T_2 = 5T$.

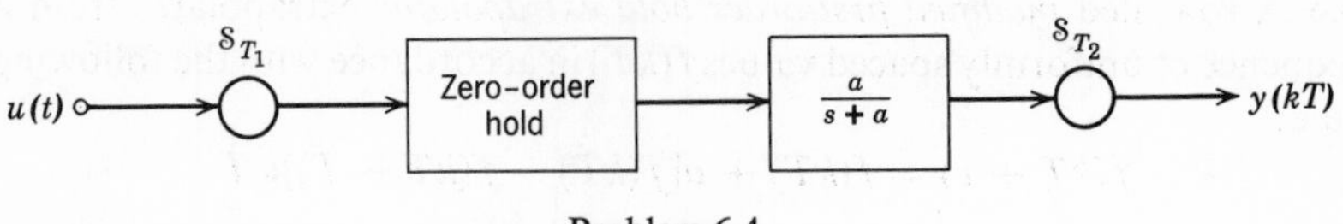

Problem 6.4

(*b*) Find $Y(z)$ for the case where $T_1 = 5T$ and $T_2 = T$.

Note:

$$e^{-aT} = \tfrac{1}{2}$$

$$e^{-bT} = \tfrac{1}{3}$$

6.5 An impulse sequence u is impressed via a zero-order-hold type of extrapolator on a system whose transfer function $H(s)$ is as follows:

$$H(s) = \frac{9}{s(s+3)}$$

(*a*) Find the z transform of the system, including the zero-order hold, given that $T = \frac{1}{3}$.

(*b*) Given that $u(kT) = e^{-6Tk}$, find $y(kT)$ in closed form.

(*c*) Determine the Laplace transform of y.

(*d*) Obtain a closed expression for $y(t)$ in the interval $8T \leq t < 9T$.

6.6 An impulse sequence is supplied via an "exponential-hold" extrapolator (i.e., via one that attenuates exponentially with time) to a plant. The extrapolator time constant is 2 sec. The transfer function of the plant is given by

$$H(s) = \frac{4}{s+3}$$

Let $u(kT)$ be the samples of the input and $y(kT)$ be those of the output.

(*a*) Determine the z transform of the entire system if $T = 0.5$.

(*b*) If $u(kT) = 2e^{-5Tk}$, find $y(kT)$ in closed form.

(*c*) Find $y(kT/2)$ in closed form.

(*d*) Find the complete output $y(t)$ for the interval $5T < t < 6T$.

CHAPTER 7

7.1 Determine the Laplace transform of the continuous output, $Y(s)$, of the system shown in the accompanying figure. The box marked S_T represents a sampler (delta-function modulator). All samplers operate in precise synchronism. The boxes marked H represent the transfer functions of continuous-time system components.

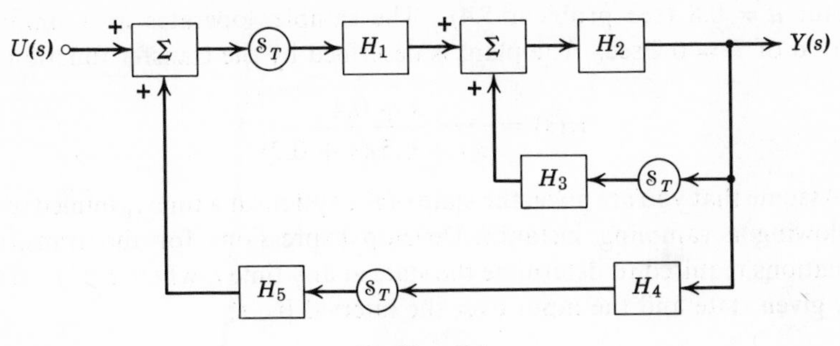

Problem 7.1

7.2 Determine $Y(s)$ for the system in the accompanying figure. Assume that the transfer functions marked H represent continuous-time subsystems and the ones marked D represent discrete-time subsystems.

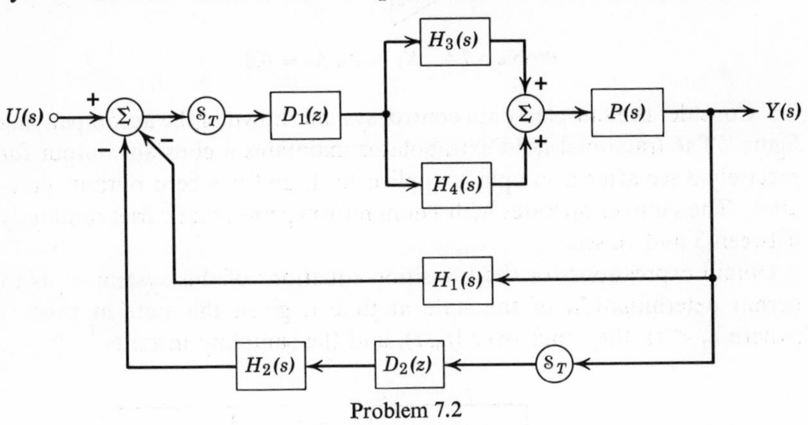

Problem 7.2

7.3 Determine $Y(s)$ for the system shown in the accompanying figure.

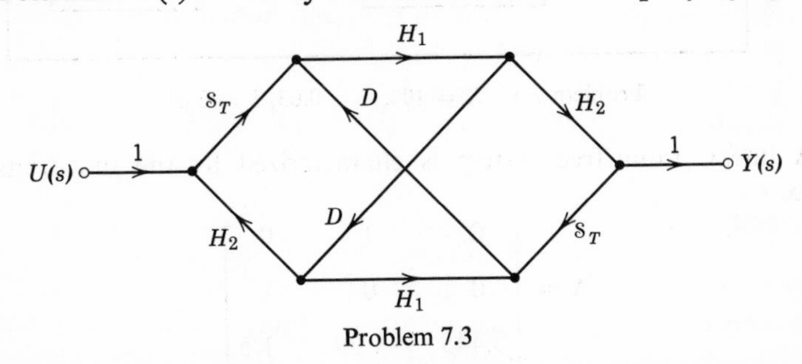

Problem 7.3

7.4 The sampled-data control system shown in the accompanying figure contains a modified first-order hold extrapolator $J^{1,a}$ with a modification factor $a = 0.8$ (see problem 5.4). The sampler operates at a uniform period of $T = 0.2$ sec. The plant is described by the transfer function

$$H(s) = \frac{s + 0.4}{s(s + 0.5)(s + 0.2)}$$

Assume that you are given the state of the system at a time t_0 immediately following a sampling instant. Develop expressions for the transition equations required to determine the state at any time t, where $t > t_0$, from the given state and the input over the interval $[t_0, t)$

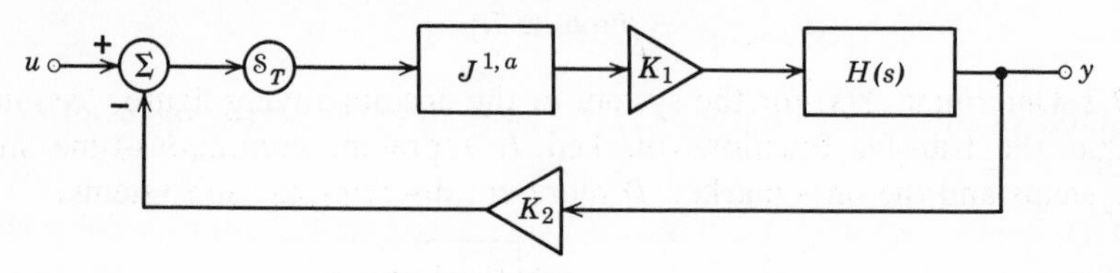

Problem 7.4 $K_1 = 20$, $K_2 = 0.8$.

7.5 Consider the sampled-data control system shown in the accompanying figure. The fractional-hold extrapolator maintains a constant output for precisely 3 sec after a sample is applied to it, and has zero output thereafter. The sampler operates with a nonuniform period that varies randomly between 3 and 10 sec.

Obtain expressions for the transition equations of this system so as to permit determination of the state at time t, given the state at time t_0 (where $t_0 < t$), the input over $[t_0, t)$, and the sampling instants in $[t_0, t]$.

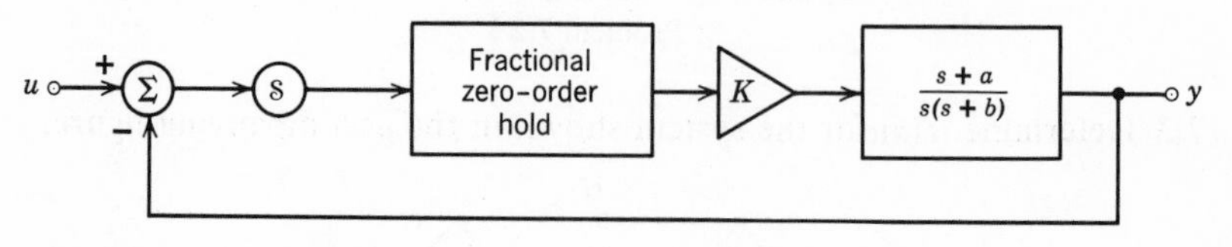

Problem 7.5 $K = 100$, $a = 0.03$, $b = 0.01$.

7.6 A linear, input-free system is characterized by the unit-transition matrix

$$\mathbf{A} = \begin{bmatrix} 0 & 1 & 0 \\ 0 & 0 & 1 \\ 0.445 & -1.39 & 1.5 \end{bmatrix}$$

Determine whether or not this system is stable by applying Liapunov's method (i.e., by using a Liapunov function).

7.7 Using the method of Liapunov, determine the necessary and sufficient conditions on the parameters a_0, a_1, and a_2 so that the system described by the difference equation

$$y(k + 3) + a_2 y(k + 2) + a_1 y(k + 1) + a_0 y(k) = 0$$

will be asymptotically stable.

7.8 Using the modified Schur-Cohn criterion, determine whether a system possessing the characteristic equation

$$8z^4 - 20z^3 + 25z^2 - 10z + 2 = 0$$

is stable or unstable. Verify the result by finding the actual values of the roots.

7.9 Determine whether the following characteristic equations belong to stable or unstable systems.

(*a*) $12z^6 + 6z^5 + 20z^4 + 3z^3 + 6z^2 = 0$
(*b*) $12z^5 - 12z^4 + 17z^3 - 9z^2 - z + 1 = 0$
(*c*) $12z^5 + 8z^4 + 5z^3 + 2z^2 + 3z = 0$
(*d*) $108z^6 - 210z^5 + 184z^4 - 81z^3 + 16z^2 - z = 0$

7.10 Develop a procedure for determining whether the equation

$$a_n x^n + a_{n-1} x^{n-1} + \cdots + a_1 x + a_0 = 0$$

(where x is a complex variable, $x = u + jv$) has any roots which lie within the circle defined by the equation

$$4v^2 + 4u^2 - 24u + 11 = 0$$

in the complex x plane.

CHAPTER 8

8.1 (*a*) Find $R_x(m)$, given that

$$S_x(z) = \frac{-0.45z}{z^2 - 2.05z + 1}$$

(*b*) If (*a*) describes the input process of a system whose weighting sequence is $(0.6)^{k-1}$ for integers $k \geq 1$, what is the autocorrelation sequence of the system output process?

(*c*) What is the mean-square value of the output?

8.2 Determine the spectral density function $S_x(z)$ for the autocorrelation sequence

$$R_x(m) = A(1 - 0.01\,|m|) \quad \text{for} \quad |m| \le 100$$
$$= 0 \quad \text{for} \quad |m| > 100$$

Indicate the region of convergence, if any.

8.3 Consider the two discrete-time systems shown in the accompanying figure. The systems are identified by their respective weighting sequences f and h. The corresponding transfer functions are $F(z)$ and $H(z)$.

(*a*) Assume that you are given the crosscorrelation sequence $R_{xu}(m)$ (where m ranges over all integers). Derive an expression for the cross-correlation sequence $R_{yv}(m)$.

(*b*) Using the result of (*a*), obtain an expression for the cross-spectral density $S_{yv}(z)$.

(*c*) Assume now that in addition to $R_{xu}(m)$, you are also given $S_x(z)$. Obtain an expression for $S_v(z)$.

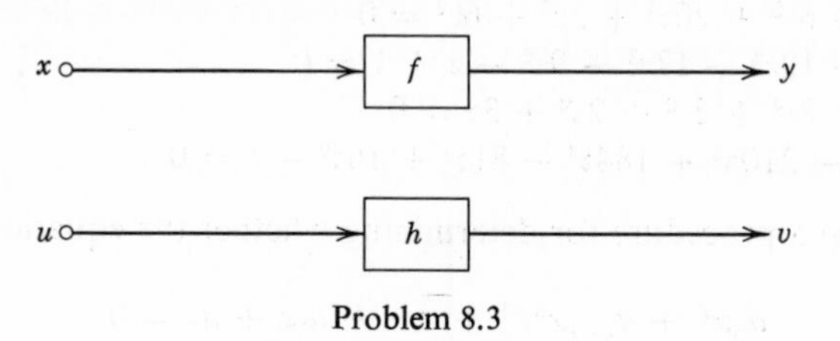

Problem 8.3

8.4 Consider the two discrete-time systems of Problem 8.3, but now assume that you are given only the cross-spectral density $S_{vy}(z)$ and are told that the spectral density $S_u(z) = 1$. Derive an expression for the spectral density $S_x(z)$ in terms of the given spectral densities and the system transfer functions.

8.5 Determine the transfer function of an optimum discrete-time predicting circuit that will "predict" the input for one time interval; that is, the output of the circuit at $k = n$ is to resemble (with minimum mean-square error) the input at $k = n + 1$. The autocorrelation sequence of the input is given by $R_x(m) = A^2 b^{-2c|m|}$.

8.6 The input f to a discrete-time system consists of a message u of spectral density $S_u(z) = -6z/(z^2 - 2.5z + 1)$ plus additive noise of constant spectral density 0.4. The message and noise are uncorrelated. The system has the transfer function $H(z) = 0.5z/(z - 0.8)$.

(*a*) What is the transfer function of the compensating system which must be placed in cascade with $H(z)$ if the output of the resultant system is to approximate the signal with minimum mean-square error?

(*b*) What is the minimum value of the mean-square error?

(*c*) What is the value of the mean square error *without* the compensating system?

8.7 Two players, A and B, toss perfectly balanced coins. If the coins show the same face (i.e., "match"), A wins B's coin. If the coins do not match, B wins A's coin. Both start with three coins. The game ends when one player has no coins left.

(*a*) What is the probability that the game ends on the nth toss?

(*b*) What is the average number of tosses required to end the game?

8.8 Solve problem 8.7 for the case where A has two coins and B has three coins.

8.9 At periodic intervals a positive or negative pulse is received by a modulo-3 counter. Negative pulses are ignored, but for each received positive pulse, the counter advances by "one," thereby going progressively from 0 to 1, from 1 to 2, and then back to 0 to repeat the cycle. There are no "carry's". The probability that any received pulse is positive is equal to $\frac{1}{2}$. The counter is started with a count of "0." What is the probability that the count will be "0" after 10 pulses have been received?

8.10 A Markov system is characterized by the transition matrix

$$\mathbf{Q} = \begin{bmatrix} \frac{1}{2} & \frac{1}{3} & 0 \\ \frac{1}{2} & 0 & \frac{1}{5} \\ 0 & \frac{2}{3} & \frac{4}{5} \end{bmatrix}$$

The cost vector is

$$\mathbf{c} = \begin{bmatrix} 3 \\ 2 \\ 1 \end{bmatrix}$$

(*a*) Determine the total cost accumulated by the system over k transitions, starting from state 1.

(*b*) As k becomes very large, what is the mean increase in cost per transition?

(*c*) What would be the difference in mean total cost as k becomes very large if, instead of starting in state 1, we had started in state 2?

8.11 A two-state Markov system is characterized by the transition probability matrix:

$$\mathbf{Q} = \begin{bmatrix} \frac{1}{4} & \frac{2}{3} \\ \frac{3}{4} & \frac{1}{3} \end{bmatrix}$$

The system incurs a transition cost as it moves from state to state in accordance with the transition-cost matrix:

$$\mathbf{D} = \begin{bmatrix} 12 & 15 \\ 8 & -12 \end{bmatrix}$$

that is, the transition from i to j, governed by the probability q_{ji}, is accompanied by a cost d_{ji}.

What is the mean total cost accumulated by the system if it starts in state 2 and makes 34 transitions?

8.12 A two-state ergodic system can be controlled by means of three possible input values, u_1, u_2, and u_3. The respective transition probability matrices and state cost vectors associated with these inputs are

$$\mathbf{Q}_1 = \begin{bmatrix} 0.20 & 0.5 \\ 0.8 & 0.5 \end{bmatrix} \qquad \mathbf{c}^1 = \begin{bmatrix} 16 \\ 6 \end{bmatrix}$$

$$\mathbf{Q}_2 = \begin{bmatrix} 0.25 & 0.2 \\ 0.75 & 0.8 \end{bmatrix} \qquad \mathbf{c}^2 = \begin{bmatrix} 15 \\ 8 \end{bmatrix}$$

$$\mathbf{Q}_3 = \begin{bmatrix} 0.833 & 0.1 \\ 0.167 & 0.9 \end{bmatrix} \qquad \mathbf{c}^3 = \begin{bmatrix} 12 \\ 9 \end{bmatrix}$$

(*a*) The system is started in state 1 and operated under an optimum policy. What will be the accumulated cost after four transitions?

(*b*) What would have been the cost in (*a*) if the system had been started in state 2 instead of state 1?

8.13 Consider the system of Problem 8.12. Assume an initial policy of $\mathbf{d}^1 = \begin{bmatrix} u_3 \\ u_1 \end{bmatrix}$, and, using an iterative scheme, determine the optimum policy for operating the system over a large numbers of transitions. What will be the average cost per transition under the optimum policy?

Index